AA003433

American Institute of Chemical Engineers

Sustainable Biorefineries

Topical Conference at the
2007 AIChE Annual Meeting

November 4-9, 2007
Salt Lake City, Utah, USA

Printed from e-media with permission by:

Curran Associates, Inc.
57 Morehouse Lane
Red Hook, NY 12571
www.proceedings.com

ISBN: 978-1-60423-832-7

Some format issues inherent in the e-media version may also appear in this print version.

ISBN: 978-1-60423-832-7

Copyright (2007) by the American Institute of Chemical Engineers.
All rights reserved.

For permission requests, please contact the American Institute of Chemical Engineers at the address below.

American Institute of Chemical Engineers
Proceedings
Three Park Avenue
New York, NY 10016-5991
Phone: 212-591-8100

www.aiche.org

American Institute of Chemical Engineers

Sustainable Biorefineries

TABLE OF CONTENTS

382a Biofuels and Biobased Products – Making Them Competitive and Sustainable

Robert L. Fireovid, USDA Agricultural Research Service, Room 4-2166, 5601 Sunnyside Ave, GWCC-BLTSVL, Beltsville, MD 20705

There's a lot of press about biofuels. But there are substantial technical barriers inhibiting a truly sizable and sustainable, biobased economy. There are also practical limits and ethical tradeoffs to how much biofuel or other bioproducts can be obtained from our remaining, arable lands.

This presentation will provide an overview of these barriers and the USDA research underway to overcome them.

382b Strategic Assessment of Prospective Biomass-to-Biofuel (B2B) Conversion Processes

Gregory Stephanopoulos, Department of Chemical Engineering, Massachusetts Institute of Technology, 77 Massachusetts Ave, 56-469C, Cambridge, MA 02139

Economic and geopolitical factors (high oil prices, environmental concerns and oil supply instability) have been prompting policy makers to pay increasing attention on renewable resources for biofuel production. This is not the first time that biomass is considered as a key feedstock for the sustainable production of a significant fraction on the total liquid fuel demand. However, current efforts for biomass-to-biofuel (B2B) conversion are characterized by renewed optimism which is rooted on, (a) the advent of the New Biology (embodied in genomic technologies and unparalleled capabilities to modulate gene expression), and, (b) rapid advances in Metabolic Engineering, the platform of technologies available for microbe engineering and strain improvement that fuses core principles of chemical reaction engineering and applied molecular biology for the design and optimization of metabolic pathways.

The key requirements of a future cost-effective biorefinery designed for liquid fuel production from biomass are, (a) a sustainable supply of biomass feedstock, (b) efficient processes for the deconstruction of biomass and release of fermentable simple sugars, and, (c) optimal microbial biocatalysts capable of converting these sugars to the biofuel of choice at high rates, yields and final concentrations. Despite significant progress in advancing the specific figures of merit in each of the above critical areas, significant challenges remain before an overall optimal B2B conversion bioprocess can be configured. This talk will identify and discuss the key limiting factors and critical challenges in each of the above areas that must be overcome before realizing a cost-competitive biofuel production process. Particular attention will be focused on schemes for sustainable biomass supply at scale, existing and emerging methods of biomass degradation, and new directions for microbe engineering for the production of ethanol as well as other future biofuels. The latter encompass a variety of options each of which poses its own specific requirements in terms of pathway construction, strain optimization, tolerance to toxic substrates and products, and overall process configuration. Metabolic engineering emerges as the key technology in meeting some of these objectives and current successes in technology development through strain improvement increases the optimism for eventual realization of cost-effective biorefineries embodying B2B conversion bioprocesses.

Sustainable Production of Liquid Transportation Fuels from Biomass - from Fields to Wheels

Paul Bryan
VP – Technology
Chevron Biofuels

AIChE / Salt Lake City – November 2007

© Chevron 2007

Chevron

Acknowledgement

Dr. John D. Wind
Senior Planning Engineer
Chevron Energy Technology Company

Fulbright Senior Specialist
Murdoch University
Perth, Western Australia

July – September 2007

"Biofuels Prospects in Australia"

© Chevron 2007

Chevron

Chevron – Biofuels Business Unit

- Created in June 2006
- Objectives:
  - Build a profitable, large-scale biofuels business for Chevron to support feedstock diversification
  - Improve the performance of first-generation biofuels:
    - Quality assurance at current fuel standards
    - Develop "industrial-scale" infrastructure
  - **Develop next-generation processing technology to significantly expand the scale and choice of feedstocks and the quality of fuel products**

© Chevron 2007

Chevron Biofuels R&D Alliances

Weyerhaeuser – **Broad Corp. Alliance / Biofuels Vertical Integration**

- Only large, integrated forest product company remaining in USA; vast expertise in feedstock production, logistics, and processing.

NREL – **Only U.S. National Lab devoted to renewable energy**

- Second generation biofuels production from biologic pathways (e.g., algae); biomass-conversion technology.

C2B2 – **Colorado Center for Biofuels & Biorefining**

- Univ. of Colorado (Biotech., Process), Colorado State (Ag. Sci.) & Colorado School of Mines (Thermochemical Conversion); close connection with NREL

UC Davis – **Ag. Sciences, Biotech, Crops & Fuels Policy**

- California-based; biomass identification and development of technologies to grow, harvest and process into transportation fuels

Texas A&M – **Ag. Sciences, Energy Technology, Crop Dev.**

- Top Ag. Sciences Program; long experience with conventional fossil-based energy sources.

Georgia Tech - **Advanced Manufacturing Technologies**

- Chemical characterization of feedstock, pretreatment, fermentation of enzyme hydrolyzed biomass to bioethanol

© Chevron 2007

Global Energy Perspectives

- Drivers:
  - Global growth in energy demand, fastest in China, India and Latin America; limits on light crude oil
  - Increasing expectations surrounding climate change
  - Concerns about national energy security in the USA & elsewhere (China, India, Australia)

  - Economic development and jobs – local, regional, national, industry sectors

© Chevron 2007

As Demand Increases, Diversification of Feedstocks for Liquid Fuels will Occur

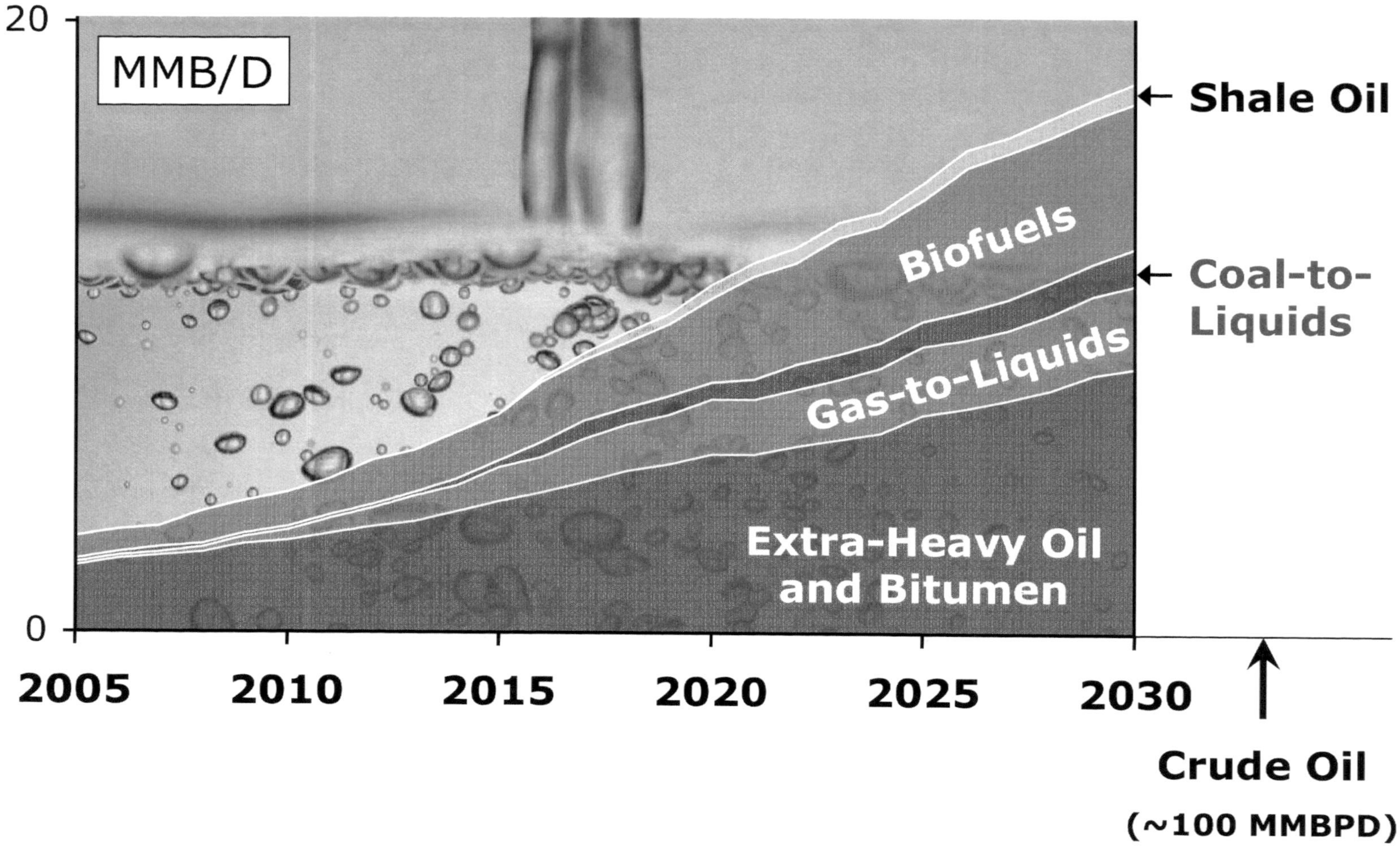

Source Data: Consultant Consensus

© Chevron 2007

Biofuels Energy Balance

Btu Spent for One Btu Available at Fuel Pump

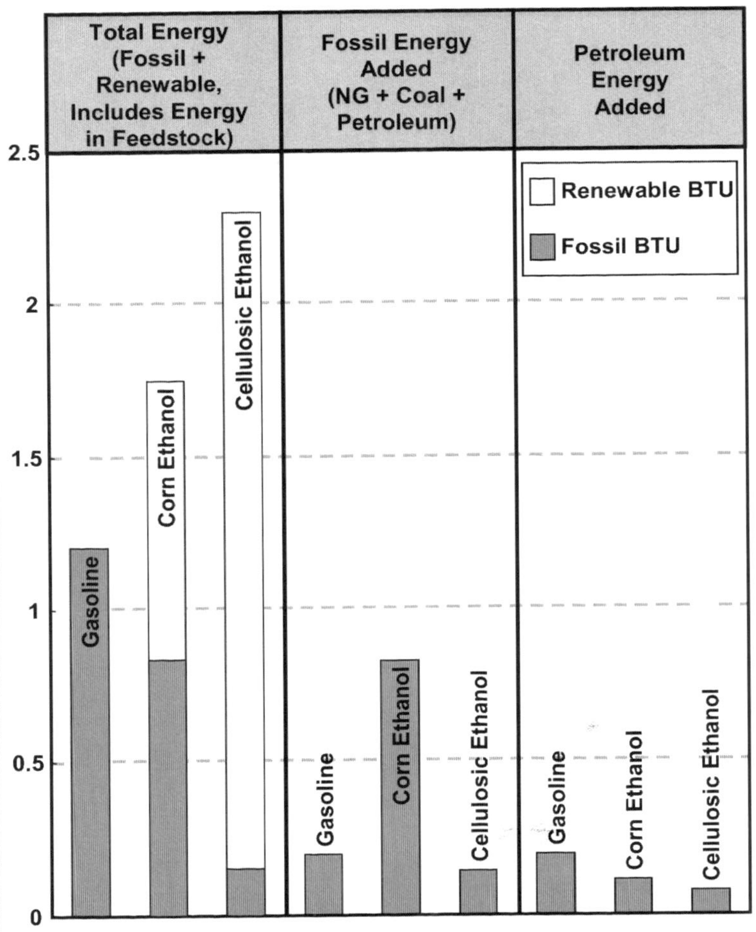

Btu Spent for One Btu Available at Fuel Pump

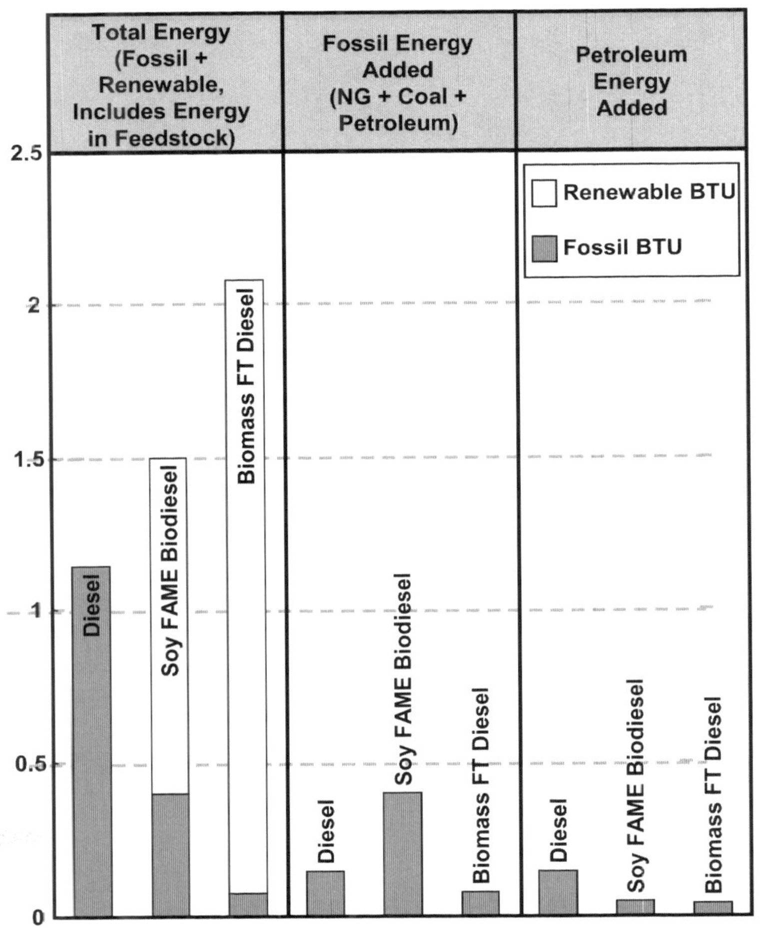

© Chevron 2007

CO2 Benefits from Biofuels

Wells / Fields to Wheels Greenhouse Gas Emissions

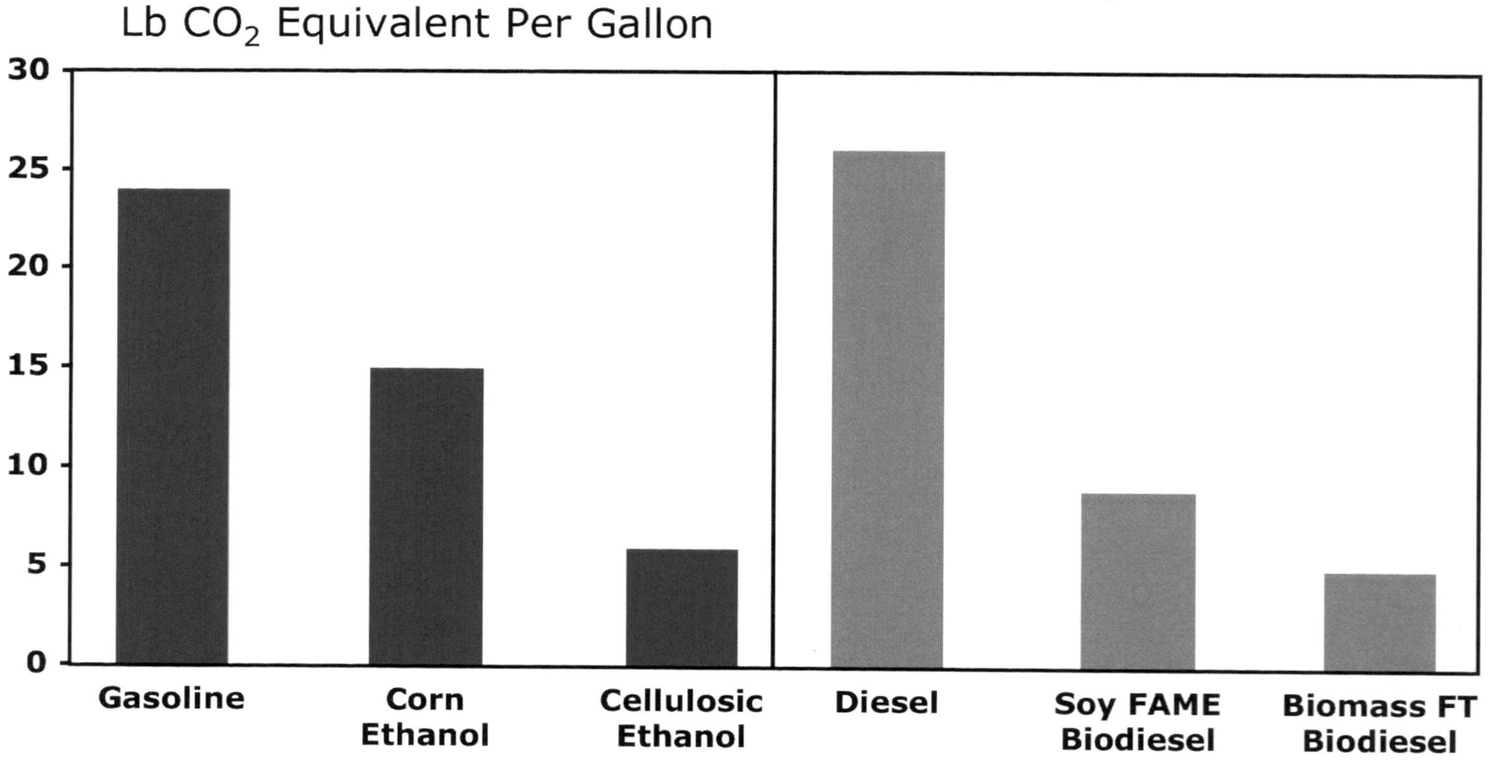

Per Gallon Gasoline Equivalent

Per Gallon Diesel Equivalent

© Chevron 2007

Source Data: Consultant Consensus

Crude Oil – Long-Term Sustainability & Short-Term Stability are both Issues

Chevron

Sovereign states – and their National Oil Companies – own most of the resources that consumers need.

Oil and Gas Reserves, Billion Barrels of Oil-Equivalent

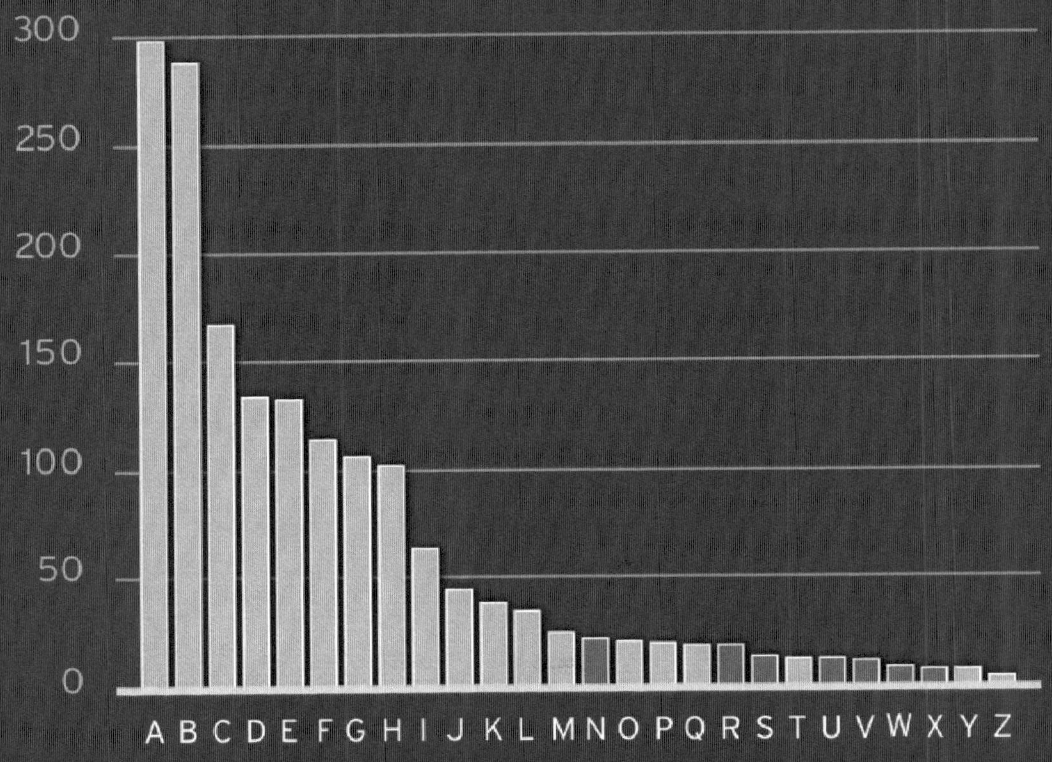

95% National Oil Companies (NOCs)

- A Saudi Aramco
- B NIOC (Iran)
- C Qatar Petroleum
- D ADNOC (UAE)
- E Iraq NOC
- F Gazprom (Russia)
- G KPC (Kuwait)
- H PDVSA (Venezuela)
- I NNPC (Nigeria)
- J NOC (Libya)
- K Sonatrach (Algeria)
- L Rosneft (Russia)
- M Petronas (Malaysia)
- O Lukoil (Russia)
- P PEMEX (Mexico)
- Q Petrochina (China)
- T Petrobras (Brazil)
- Y ONGC (India)
- Z Sinopec (China)

5% International Oil Companies (IOCs)

- N ExxonMobil
- R BP
- S Chevron
- U Royal Dutch Shell
- V Total
- W ConocoPhillips
- X ENI

© Chevron 2007

Source: Credit Suisse First Boston

Chevron

Key Issues for Biofuels

- Feedstock type and source are critical
- GHG balance varies markedly with feedstock, process technology, and prior land use
- Agricultural sustainability will depend on crops and farming practices
- Moving to a zero-waste world – every resource has energy and product value
- Need to understand balance of co-products and interaction of food and energy markets
- Compatibility with existing multi-trillion dollar fuels infrastructure will greatly impact investment required

© Chevron 2007

Biofuels Feedstocks

- Scale & seasonality issues
- Collection costs
- Crop threats - weather, land, pests
- Competition with food / feed crops:
  - Ethanol from sugar cane & corn will be limited by land availability
  - Biodiesel from soy, palm & canola will also be limited and too costly
- Future: wastes and "on-purpose" energy crops:
  - Urban wastes (paper, lumber, ...)
  - Forest slash
  - Pulp & paper waste
  - Ag waste (stover, straw, etc.)
  - Jatropha & other new oil crops
  - Switchgrass & other grasses / woods
  - Micro-algae (in ponds or reactors)
  - "Energy" crops (corn, cane, etc.)

© Chevron 2007

Desirable Feedstock Characteristics

- Grows on marginal lands (tolerance to drought, heat, waterlogging, salinity, acidity, ...)
- Minimal inputs (water, fertilizer, herbicides, ...)
- Minimal competition with indigenous food and animal-feed crops
- High yields, consistent year-to-year
- Ease of processing → biofuels
- Valuable co-products
- Easy to harvest / store

© Chevron 2007

Chevron

Agricultural GHG Emissions

- GHG Emissions from:
  - Tillage (plowing, seeding, initial cultivation)
  - Fertilizer (N, P, K), pesticide, herbicide, lime, ...
  - Irrigation
  - Harvesting (cutting, baling, ...)
  - Drying
  - Transportation
  - Metabolic emissions of CO_2, **N_2O, and CH_4**
- GHG Sequestration in:
  - Above-ground biomass
  - Root growth
  - Soil organic matter accumulation

© Chevron 2007

Choice of Crop Greatly Affects Emissions

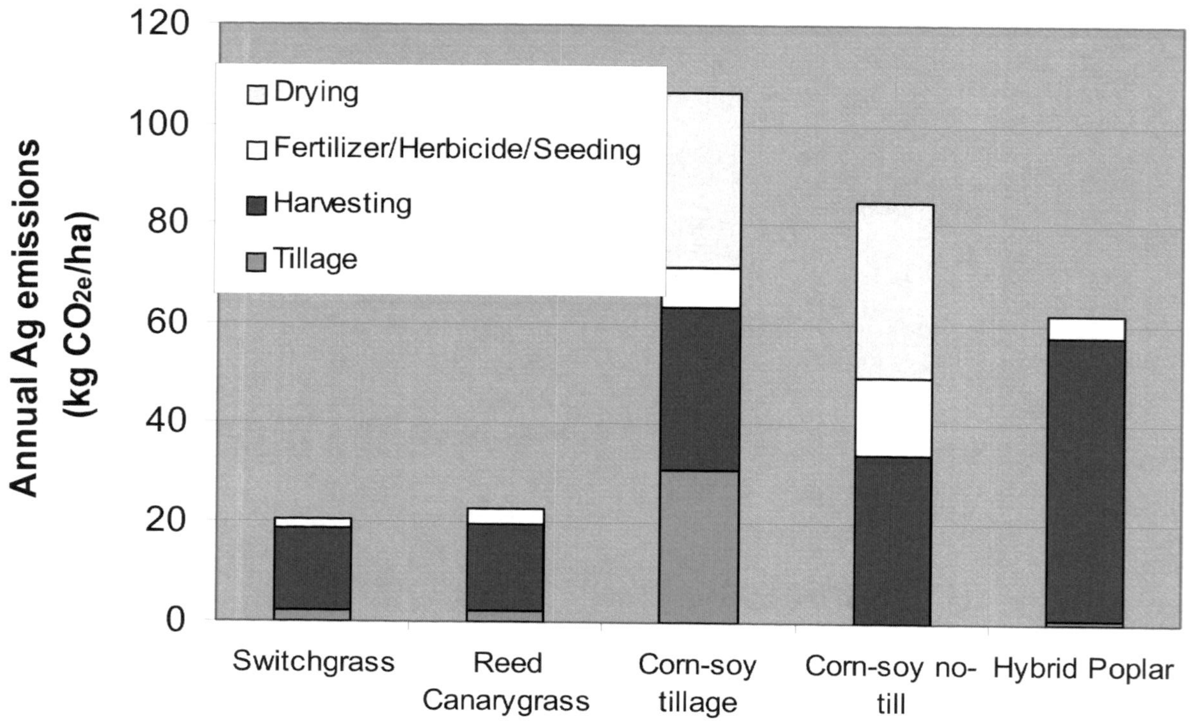

Fuel Yield: **7** **5** **6** **6** **8**
(MJ/m^2-yr)
(approximate)

Data from Adler et al, *Ecological Applications, 17(3)*, **2007**

© Chevron 2007

Conclusions

- Crop selection and farming practices have the greatest impact on sustainability
- Conversion technology has impact primarily be enabling specific feedstocks
- Scale is critical – sustainability will have little impact if the fuel source is small in scale
- Economic sustainability is important in determining the long-term viability of any biofuels value chain; this includes secondary effects on costs of other goods & services, especially food
- Biofuels are not the only answer to energy supply and sustainability, but they are a key piece of the puzzle
- There will likely be multiple biofuel solutions, each best suited to different climates and techno/socio/economic situations

© Chevron 2007

382d Second Generation Biofuels and Biochemicals: An Industry Perspective

Todd A. Werpy, Paul Bloom, and Michael Pacheco. Archer Daniels Midland, 1001 N. Brush College Rd., Decatur, IL 62521

The term biorefinery has been coined as representing the opportunity to produce multiple products from a single feedstock within a single facility. This concept has been put into practice for decades in the agricultural processing industry, particularly the corn wet milling processing industry. The basic concept for corn wet milling involves the separation of corn into four major components, corn oil, corn fiber, protein and starch. The protein, oil and fiber are generally sold into the food and feed markets. Starch is the primary product and is converted to multiple products within a single facility. The list of products includes both fuels such as ethanol, food products, and chemical intermediates. Processes for converting starch include both fermentation to products such as lactic acid and amino acids. Direct chemical conversions, including hydrogenation are employed to afford products such as sorbitol.

The biorefinery of the future will utilize both existing agricultural products such as corn but will also move to utilize new feedstocks such as corn fiber, stover and potentially energy crops such as swtichgrass. The new biorefineries will produce traditional renewable transportation fuels such as ethanol but will also likely produce new fuels that are based on thermal conversions to produce bio-derived oils. Chemical opportunities in the new biorefineries will have the potential to be expansive and include chemical products based on a variety of different carbon number platforms including C2, C3 and C4. These platforms will produce traditional drop in chemical replacements such as propylene glycol, ethylene glycol, 1,4-butanediol and a family of others. In addition, new chemical intermediates with different functionality will provide an opportunity for new growth in the chemical markets.

This presentation will discuss some of the future opportunities associated with the development of the next generation biorefineries that provide an economically viable pathway for the utilization of biomass to produce both fuels and chemicals.

382e Cellulose-to-Ethanol Technology: An Update from Iogen

Glenn D. Munkvold, Iogen Corporation, 310 Hunt Club Road East, Ottawa, ON K1V1C1, Canada

The talk will discuss the role of cellulosic ethanol in meeting targets for reductions in green house gas emissions and petroleum consumption, with particular reference to how Iogen's cellulose-to-ethanol technology can make a cellulosic ethanol industry a reality.

SURVEY OF COMMERCIAL BIOMASS GASIFIERS

During the past year I have been surveying commercially available biomass gasifiers, developed principally in Europe and Japan. At the 2007 Spring AIChE Meeting in Houston I presented a tutorial, based upon surveys I performed for two clients, in which I summarized the status of a dozen technologies that appeared to be useful for those two clients to examine more closely for the production of synthesis gas to use in chemicals manufacturing. The paper and slides for that presentation are to be found on my webpage at the University of Pittsburgh (http://www.engr.pitt.edu/chemical/facstaff/cobb.html).

Following the Spring Meeting I shifted my focus to developing a listing of biomass gasifiers that have been demonstrated and/or placed in commercial operation in projects sized greater than 7 MWe or 100 tons of biomass per day. By "biomass" I refer to wood chips, straw (including corn stover), RDF, MSW and waste plastic. I exclude sewage sludge, animal waste (including poultry liter) and lignin (blackwater). Ultimate products of the gasification plants include synthesis gas, steam (one example being for district heating) and electricity. Thus, they include both air-blown, oxygen-blown, steam-blown and indirectly heated gasifiers. I reviewed my earlier presentation, sought the advice of staff at the National Renewable Energy Laboratory, and spoke with several others with experience in this area. This exploration has yielded a list of thirteen technologies.

In alphabetical order the thirteen gasification technologies are: BGL; Choren; Enerkem (BIOSYN); Foster Wheeler ACFB; Foster Wheeler PCFB; Lurgi CFB; Lurgi Dry Ash; Primenergy; Repotec; SylvaGas; Thermoselect; TPS; and Westinghouse Plasma. Here is a brief sketch of these thirteen technologies:

BGL – 840 T/D (156 MWt) commercial plant, 2003, in Spreetal, Germany
> **Feed**: MSW, coal/biomass briquettes
> **Products**: methanol and electricity
> **Gasifier Parameters**: downward moving bed; slagging bottom; steam/oxygen-blown; 1450°C; 25 bar

Choren – 200 T/D beta plant (U.C.) in Freiberg, Germany
> **Feed**: 68,000 T/Yr of feed (50% waste wood, 50% clean wood chips)
> **Product**: synthesis gas that is converted to 16.5 million liters/year of SunFuel via Fischer Tropsch synthesis and hydrocracking
> **Gasifier Parameters**: sequential pyrolyzer (horizontal, stirred bed; 500°C; 4 bar), gasifier of pyrolysis gas (1400°C; 4 bar) and "chemical quencher" of blend of pyrolysis char and gasified pyrolysis gas (entrained bed; 900°C; 4 bar); oxygen-blown

Enerkem BIOSYN – 7 MWe (90 T/D) demonstration plant, 2003, in Ribesalbas, Spain
> **Feed**: MSW, waste biomass, RDF, plastic
> **Products**: synthesis gas, electricity
> **Gasifier Parameters**: bubbling fluid bed; air- or oxygen-blown; 1832°F; 16 atm

Foster Wheeler ACFB – 42 MWe commercial plant, 1998, Lahti, Finland
Feed: biofuels, RDF, wood waste
Product: electricity
Gasifier Parameters: circulating fluid bed (CFB); air-blown; 1000°C; 1 atm

Foster Wheeler PCFB – 6 MWe and 9 MWt (55 T/D) demonstration plant, 1999, Värnamo, Sweden
Feed: wood, RDF and straw
Product: electricity
Gasifier Parameters: CFB; air-blown; 950°C; 18 bar

Lurgi CFB – 84 MWt and 29 MWe (500 T/D) commercial plant, 2000, in Lahden, Netherlands
Feed: demolition wood, RDF
Product: electricity
Gasifier Parameters: CFB; air-blown; 900°C; 1 atm

Lurgi Dry Ash – 340 T/D (410 MW) commercial unit, 1964, in Spreetal, Germany
Feed: plastics, MSW pellets, tar-sludge pellets, contaminated wood, other wastes
Products: fuel gas, synthesis gas
Gasifier Parameters: grate-type fixed-bed down-flow dry-ash reactors; oxygen/steam-blown; 1472-2372°F; 25 atm

Primenergy – 290 T/D (1 MWe) CHP commercial plant, 2006, in Little Falls, Minnesota
Feed: waste wood
Products: electricity, steam
Gasifier Parameters: updraft gasifier; air-blown; 1 atm

Repotec – 8 MWt demonstration plant, 2001, in Güssing, Austria
Feed: wood chips
Product: 2.0 MWe and 4.5 MWt for district heating
Gasifier Parameters: coupled CFBs (biomass pyrolyzer, char combustor); steam-blown (pyrolyzer) and air-blown (combustor); 850°C (pyrolyzer)

SylvaGas – 20 MWe (400 T/D) demonstration plant (U.C.) in Forsythe, Georgia
Feed: clean biomass
Product: synthesis gas
Gasifier Parameters: coupled CFBs (biomass pyrolyzer, char combustor); steam-blown (pyrolyzer) and air-blown (combustor); 1500°F (pyrolyzer) and 1800°F (combustor); 1 atm

Thermoselect – 600 T/D commercial plant, 2005, in Kurashiki, Japan
Feed: MSW plugs
Product: synthesis gas
Gasifier Parameters: sequential biomass pyrolyzer and oxygen-blown, entrained char gasifier; 570°F (pyrolyzer) and 2,200°F (combustor)

TPS – Two 15 MWt units (110 ton/hr each) in demonstration plant, 1992, in Grève-in-Chianti, Italy

Feed: RDF pellets

Product: electricity

Gasifier Parameters: CFB; air-blown; 850-900°C; 1 atm

Westinghouse Plasma – 8 MWe (300 T/D) commercial plant, 2003, in Utashinai, Japan

Feed: MSW, ASR

Product: electricity

Gasifier Parameters: downward moving bed; plasma bottom; air-blown; 4500°F (plasma torch) and 2282°F (gas exit); 1 atm

The user of this list should define the feed, product and gasifier parameters of their project to determine which technologies possess characteristics appropriate for it.

Characterization Of A Novel Gasification Transport Reactor For Conversion Of Biomass

Larry Pearson[1], Hossein Toghiani[2], Brian M. Kauffman[1], Rebecca K. Toghiani[3], John A. Etheridge[1], R. Arunkumar[1], and Jay McCown[1]. (1) Institute for Clean Energy Technology, Mississippi State University, 205 Research Blvd, Starkville, MS 39759, (2) Chemical Engineering, Mississippi State University, 330 Swalm Circle President Dr., Mississippi State University, MS 39762, (3) Chemical Engineering, Dave C. Swalm School of Chemical Engineering, Mississippi State University, 330 Swalm, Mississippi State, MS 39762

For some years, the cost and availability of petroleum feedstocks have been a concern for industrial and manufacturing operations. With the instability currently being experienced in the world's major oil-producing regions, many companies and governmental agencies have begun looking to alternative sources as possible substitutes for petroleum-based feedstocks. Biomass has emerged as perhaps the most promising near-term option. Biomass feedstocks have been utilized for centuries to provide process heating. The use of biomass as a feedstock in industrial processes, however, requires a conversion system capable of generating hydrogen (H_2) and carbon monoxide (CO) in ratios and quantities that allow utilization of proven, petrochemical processes.

Mississippi Ethanol (ME) and the Institute of Clean Energy Technology have participated with the Department of Energy (DOE) in a comprehensive effort to develop the conversion of biomass utilizing a proprietary gasification reactor. The "Mississippi Ethanol Gasification Project", DE-FG36-02GO12025, divided tasks between operation of a pilot-scale unit and a pre-pilot unit (1/10 pilot-scale). Capacity of the pilot-scale unit was a nominal 30 tons/day of waste wood. Each unit was characterized for gas composition versus process operating parameters. In addition, the data from both units w evaluated to develop design scale-up factors.

The gasification reactor system, U.S. Patent No. 6,767,375, is of an entrained flow, indirectly heated design. Biomass, in the form of wood sawdust in this project, is introduced into the reactor by means of a proprietary feeder along with a transport gas, typically steam. In the reactor, the biomass is converted to a "syn-gas" by a series of reactions involving pyrolysis, reforming, and gasification. The composition of the product syngas is dependent on several operating parameters, including reaction temperature, residence time, and steam-to-biomass ratio.

The two reactor systems tested, pilot and 1/10 scale pre-pilot, were designed and constructed to allow comparison of data and development of scale-up factors. As such, each system consisted of a wood feed system, a gasification reactor, syngas cooling, ash removal, and a flare system. Superheated steam was produced in both units utilizing a syngas/steam cross exchanger. To simplify operation, heat for the gasification reactor was provided by natural gas fired burners. Sampling of the syngas was performed immediately after ash removal. Each unit was fitted with a PC-based supervisory control and data acquisition system complete with alarm management, process visualization, and data archival capabilities.

The pre-pilot and pilot unit systems were operated across a range of operating conditions. The biomass feed to both units consisted of sawdust from a local cabinet shop which was dried, as necessary, and sized to approximately 1/8 inch topsize. Syngas compositions from this testing showed hydrogen concentrations from 20 to 45 mol%, and corresponding carbon monoxide concentrations of 55 to 25 mol%. Methane concentrations were less than 10 mol% for all test runs.

Data from testing were analyzed to characterize the pilot and 1/10 scale pre-pilot systems. Since one of the objectives of this testing was to identify system product gas compositions with the intent of developing a correlation between the pilot and pre-pilot unit data, a concentrated effort was undertaken to characterize the product compositions utilizing easily defined operational parameters. Two major factors affecting the product gas composition in the ME gasification system were the reactor temperature and the reactor residence time. A reactor equivalent temperature, T(eq) was constructed which utilized 1) the reactor exit temperature and 2) the reactor steam feed temperature. Similarly, a reactor equivalent residence time, Res(eq), was constructed, which made use of 1) the total feed rate to the reactor, 2) an equivalent density, and 3) the steam/wood ratio. Finally, product gas compositions were compared to an overall equivalent number calculated as a simple combination of T(eq) and Res(eq), (ie T(eq) x Res(eq)). The results of these comparisons were fitted using a polynomial equation.

One of the primary goals of this research was to utilize the characterization data gathered from both the pre-pilot and pilot units to determine relationships which could be used to scale-up the design to larger capacities. This would allow for small scale testing which could then be related to a pilot scale unit and larger. As such, the characterization data from both units were compared and related as a function of gasification reactor geometry. The correlation allowed generation of a scale-up curve for reactor design.

In order to replace, or supplement, petroleum feedstocks in industrial processes, such as the production of diesel or alcohol, a key requirement is the production of an appropriate hydrogen-to-carbon monoxide ratio. This testing has shown that the Mississippi Ethanol Gasification Reactor System is capable of utilizing biomass to produce a gas with compositions suitable for diesel or alcohol production. In addition, scale-up correlations have been developed to allow for the design of larger, commercial scale facilities.

400c Gasification Of Biomass Constituents By Supercritical Water In A Microchannel Reactor

Aaron Goodwin and Greg Rorrer. Chemical Engineering, Oregon State University, Corvallis, OR 97331

Lignocellulosic biomass is a renewable, CO_2 neutral feedstock for the production of liquid and gaseous fuels. Lignocellulosic biomass consists of cellulose, hemicellulose and lignin. Cellulose is a polymer of the six-carbon sugar glucose, hemicellulose is a polymer of five-carbon sugar xylose, and lignin is a polymer of phenylpropane units. In general, monomeric carbohydrate compounds such as glucose and xylose can be non-catalytically gasified in supercritical water at a temperature greater than 600°C and pressure of 240 bar to yield a gas stream primarily composed of H_2, CO_2, and CH_4. However, due to the endothermic nature of the gasification reaction, and relatively low rates of heat transfer in conventional tubular reactors, the residence time for complete gasification of carbohydrates is at least 20 seconds. Microchannel reactor technology offers significant promise to intensify the gasification of biomass constituents by increasing the rate of heat transfer to the reaction. In this study, we considered gasification of soluble biomass constituents by supercritical water in a microchannel reactor. The novel stainless-steel microchannel reactor consisted of serpentine parallel array of 75 μm by 500 μm rectangular channels. This reactor greatly intensified the gasification reaction in supercritical water. For example, at 750°C and 250 bar, a 0.1 molar solution of glucose was gasified to 53% H_2, 35% CO_2, 10% CH_4, and 0.5% CO in less than 2.5 seconds. The liquid effluent was clean and contained only trace amounts of acetic acid. Ongoing work using other biomass constituents, including xylose, monomeric lignin precursors, and an alkali-soluble hemicellulose/lignin stream isolated by NaOH pretreatment of grass straw, is also described.

400d Utilization Of Biorenewable Materials For Producing Energy, Chemicals, And Fuels Via Fast Pyrolysis Approach

Justinus A. Satrio, Center for Sustainable Environmental Technology, Iowa State University, 273 Metals Development, Ames, IA 50011 and Robert C. Brown, Center for Sustainable Environmental Technologies, Iowa State University, 286 Metals Development, Ames, IA 50011.

Biomass materials, such as switchgrass and cornstovers, are very potential to be used as renewable resources for producing energy, chemicals, and liquid fuels. The utilization of biomass, however, is limited by the cost of transporting. In addition to the location for collecting biomass, which is generally far away from the production facilities, biomass has a low energy density thus it is expensive to transport. Fast pyrolysis has been seen as a promising approach to improve the efficiency of biomass utilization by reducing the cost of transportation. In this approach biorenewable materials, such as cornstovers, are collected and preprocessed locally to yield energy-rich biooil and porous char via fast pyrolysis. The char can be injected back to the soil or used for producing heat. The biooil, which can be thought as densified biomass (density is 1280 kg/m3 compared to approximately 160 kg/m3 for original biomass), is transported by tanker trucks to a centralized facility where it is used as feedstock to produce energy, chemicals, and/or liquid fuels. The potential benefits of this system include: 1). Efficient and economically profitable utilization of abundant biomass which is otherwise considered waste, 2) Reduction in the consumption of expensive fossil fuel and natural gas which potentially reduces the emission of CO2. The use of biooil is not without problem, however. Biooil contains a good number of functional groups, such as carbonyl, carboxyl, phenolic, and hydroxyl, which renders the oils reactive and thermally unstable. In addition, often time bio-oil needs to be separated into components/fractions to be used as feedstock for certain applications. Iowa State University has developed a fast pyrolysis process for producing bio-oil from biorenewable materials, such as cornstovers and switchgrass. The fast pyrolysis unit consists of a 16.2 cm diameter fluidized bed reactor, a burner to externally heat the reactor, a two-stage auger to feed the solid, two cyclones to remove particulate matter, and a vapor-condensing system consisting of four condensers and an electrostatic precipitator. The condenser system enables collection of high water content biooil fractions of bio-oil with very low lignin content and high lignin fractions with low water content. Effects of reaction conditions and condenser temperatures on the bio-oil product distribution will be discussed. Also discussed are factors that affect bio-oil product stability and steps that can be taken to improve bio-oil stability. Finally, the potential utilization processes of these fractions for producing energy, chemicals, and fuels will be addressed.

400e A Comparative Study Of Rapid Pyrolysis Oils From Woody Biomass And Poultry Litter

Foster Agblevor and Frederick Teye. Virginia Tech, 212 Seitz Hall, Biological Systems Engineering Department, Blacksburg, VA 24061

Bio-oils produced from the rapid pyrolysis of poultry litter and hardwoods were observed to have considerable differences in their higher heating values, nitrogen contents, sulfur contents, and viscosities. In order to develop suitable applications of these potential biorefinery feedstocks, it is essential to characterized bio-oils with respect to their physico-chemical properties. Thus, bio-oils produced from the rapid pyrolysis of hardwood bedding material, broiler litter, and turkey litter were fractionated into acidic, basic, neutral and hydrocarbon fractions. The fractions were then characterized by gas chromatography, gas chromatography/mass spectrometry, Fourier Transform infrared spectrometry, 13C-NMR, elemental analysis, and higher heating values (HHV). The bio-oils from the poultry litter had higher viscosities, nitrogen content, and HHVs than the corresponding samples from the hardwood. The hydrocarbon and neutral compounds content of the poultry litter oils were ten times higher than the corresponding hardwood biooils. Some oil fractions had very high levels of polymeric material and were very difficult to analyze by any of the above methods. There is a need to develop other characterization tools that will account for all fractions and analyze the bio-oils without necessarily fractionating them.

400f Catalytic Upgrading Of Bio-Oil Vapor

Vamshi K. Guda[1], Hossein Toghiani[2], Charles U. Pittman Jr.[1], Amit C. Gujar[2], Ravi Challa[3], and Phil Steele[4]. (1) Department of Chemistry, Mississippi State University, Mississippi State University, MS 39762, (2) Dave C. Swalm School of Chemical Engineering, Mississippi State University, Mississippi State University, MS 39762, (3) Chemical Engineering, Mississippi State University, Mississippi State University, MS 39762, (4) Forest Products Department, Mississippi State University, Mississippi State, MS 39762

Fast pyrolysis of biomass generates a complex vapor phase containing a variety of compounds, some 20 % in the form of fine aerosols. Before packed bed reactor studies could begin, it was imperative that vapor phase batch reactor studies of the catalytic upgrading of this stream be undertaken to provide a means for rapid catalyst screening under readily controlled operating conditions. A Parr batch reactor was employed in the batch studies. Bio-oil is an extremely complex mixture. Therefore, initial studies have focused on three model bio-oil compounds and also a bio-oil simulant mixture comprised of ten compounds. These ten compounds were chosen so that important functional groups present in the bio-oil were well-represented. Computer simulation was used to determine reactor charge required to achieve the vapor phase at the desired operating temperature and pressure. The impact of catalyst composition, reactor temperature, reactor pressure, and reaction time on the conversion and product distribution for each of the model bio-oil compounds and for the bio-oil simulant. Catalysts employed were acid- and base-cracking catalysts, hydrodeoxygenation catalyst and hydrogenation catalyst. Catalysts were characterized using BET, XRD and TEM.

404a Aqueous-Phase Hydrogenation Of Organic Acids On Monometallic Supported Catalysts: A Combined Experimental And Theoretical Study

Ye Xu, Center for Nanophase Materials Sciences, and Chemical Sciences Division, Oak Ridge National Laboratory, P.O. Box 2008, MS-6493, Oak Ridge, TN 37831-6493, Hakan O. Olcay, Department of Chemical Engineering, University of Massachusetts, Amherst, Goessmann Laboratory, N. Pleasant Street, Amherst, MA 01003, and George Huber, Chemical Engineering, University of Massachusetts-Amherst, 159 Goessmann Lab, 686 North Pleasant St., Amherst, MA 01003.

Depletion of fossil fuel combined with environmental and political concerns with fossil fuels have accelerated research on technologies to convert biomass into liquid fuels, i.e biofuels. Aqueous-phase hydrogenation (APH) reactions are crucial for biofuel production including: ethanol production by hydrogenation of fermentation products (organic acids) [1], gasoline production by hydrogenation of pyrolysis-oils [2], and alkane production by aqueous-phase dehydration/hydrogenation of carbohydrates [3]. APH reactions involve hydrogenating a range of functionalities, including acids (C-OOH bonds), sugars (C-O-C bonds), alkenes (C=C bonds) and aldehydes (C=O bonds). The objective of this project is to develop a fundamental understanding for aqueous-phase hydrogenations of organic acids by a combined theoretical and experimental approach.

We are in the process of measuring the catalytic activity for aqueous-phase hydrogenation of acetic acid with supported monometallic Ru, Pt, Pd, Ni and Cu based catalysts in a fixed bed reactor. The Ru catalyst is significantly more active than Pt. The reaction is first order with respect to H_2 for Ru/Carbon with an apparent activation energy of 31 kJ/mol. The reaction is 0.6 order with respect to H_2 for Pt-based catalysts with an apparent activation energy of 33 kJ/mol.

These experimental data are combined with density functional theory (DFT) calculations to offer us insight into why some monometallic catalysts are more active. We have calculated the energetics and kinetics of probable elementary steps for acetic acid hydrogenation on Ru(0001). The results are compared with previous DFT calculations for hydrogenation of acetic acid on Pt, Pd and Cu [4-6]. With the additional information afforded by theory we are attempting to identify the physical origin of the difference in the activities of these monometallic catalysts. This study will provide valuable insight for designing a more versatile APH catalyst in the future, which will probably be bi-metallic.

[1] T. Eggeman, D. Verser, Recovery of organic acids from fermentation broths, Appl. Biochem. Biotechnol. **2005**, *121-124*, 605.

[2] G. W. Huber, S. Iborra, A. Corma, Synthesis of transportation fuels from biomass: chemistry, catalysts and engineering, Chem. Rev. **2006**, *106*, 4044.

[3] G. W. Huber, J. A. Dumesic, An overview of aqueous-phase catalytic processes for production of

hydrogen and alkanes in a biorefinery, Catal. Today **2006**, *111*, 119.

[4] R. Alcala, J. W. Shabaker, G. W. Huber, M. A. Sanchez-Castillo, J. A. Dumesic, Experimental and DFT studies of the conversion of ethanol and acetic acid on PtSn-based catalysts, J. Phys. Chem. B **2005**, *109*, 2074.

[5] V. Pallassana, M. Neurock, Reaction paths in the hydrogenolysis of acetic acid to ethanol over Pd (111), Re(0001), and PdRe alloys, J. Catal. **2002**, *209*, 289.

[6] M. A. Natal Santiago, M. A. Sanchez-Castillo, R. D. Cortright, J. A. Dumesic, Catalytic reduction of acetic acid, methyl acetate, and ethyl acetate over silica-supported copper, J. Catal. **2000**, *193*, 16.

A Kinetic and Mass Transfer Model for Glycerol Hydrogenation in a Trickle Bed Reactor

Yaoyan Xi[a], Johnathan E.Holladay[b], John G. Frye[b], Aaron A. Oberg[b], James E. Jackson[a]
Dennis J. Miller[a]
[a]Chemical Engineering and Materials Science Department, Michigan State University, East Lansing, MI 48824, USA
[b]Pacific Northwest National Laboratory, PO Box 999, Richland, WA 99352, USA

Abstract

Glycerol hydrogenolysis to propylene glycol was carried out in a trickle bed reactor over a broad range of reaction conditions. Outlet glycerol conversion, selectivity to PG, and temperature profiles along the trickle bed were collected experimentally to characterize the system. A detailed model of glycerol hydrogenolysis has been developed that includes a mechanistically based kinetic rate expression, energy transport, mass transport through the liquid phase, and partial wetting of the trickle bed. Optimal kinetic parameters based on all collected data from the trickle bed system were determined via regression analysis. Model predictions agree with experimental data and accurately predict trends in reactor performance with liquid flow rate, temperature, hydrogen pressure, and base promoter concentration. The kinetic rate expression represents truly intrinsic kinetics and is compared with batch reactor data to unify reactions in both environments. The model is thus a useful tool for predicting laboratory reactor performance and for preliminary design of commercial-scale trickle bed systems.

Key words: Hydrogenolysis, modeling, glycerol, propylene glycol, trickle bed reactor

Introduction

Multiphase reactors are widely used in the production of petroleum-based fuels, commodity and specialty chemicals, pharmaceuticals, and polymers[1-5]. Trickle bed reactors are especially favored thanks to easy installation, flexible configurations and operation conditions, and good mass transfer between phases. They are widely employed in hydro treating processes such as hydrogenolysis, hydrodesulfurization, hydrocracking and hydrorefining.

Studies that correlate trickle bed reactor performance with operating conditions have been conducted. Larachi et al[6] investigated the gas-liquid interfacial mass transfer in trickle bed reactors at elevated pressure, and proposed a rigorous thermodynamic model accounting for liquid and gas nonidealities along with pressure, gas and liquid superficial velocities, liquid viscosity, and packing size. Iliuta et al [7] studied the fluid dynamics and the gas-liquid mass transfer characteristics of trickle-bed reactors. A more quantitative model was summarized by Al-Dahhan et al[2] taking into account flow regime transitions, liquid holdup, pressure drop, gas-liquid interfacial area, mass transfer coefficients, and catalyst wetting efficiency in both liquid-limited and gas-limited scenarios. Huang et al [8] studied the heat transfer characteristics of trickle bed reactors on the laboratory scale and determined the four factors to influence the local fluid-solid convective heat transfer coefficient. A correlation for catalyst wetting efficiency in elevated pressure trickle bed reactors was developed by Al-Dahhan [9] to correlate laboratory and pilot scale data. Tronconi et al [10] developed a mathematical model to characterize carbohydrate hydrogenolysis.

Despite the studies mentioned above, little work has been done to integrate mass transfer and the energy balance into a trickle bed model along with operating parameters. Since trickle bed reactor modeling is such a powerful tool, enabling a better understanding of the glycerol hydrogenolysis process in terms of trickle bed geometry, gas-liquid-solid mass and heat transfer, intrinsic reaction kinetics, and process operational conditions, this paper will mainly focus on the establishment and development of our model based on experimental data.

All the data were derived from glycerol hydrogenolysis using a 40 cm^3 trickle bed reactor loaded with a proprietary carbon-supported mixed metal catalyst (Figure 1). The catalyst bed was sandwiched between a layer of 2 mm diameter glass beads at the bottom of the bed and a layer of 2 mm diameter stainless steel beads at the top of the bed that facilitate liquid distribution and preheating prior to reaction. The trickle bed reactor is heated by a recirculating silicon oil stream flowing through a jacket surrounding the reactor and controlled by a Julabo programmable recirculating bath. A thermowell was installed along the centerline of the trickle bed axis through the bottom of the catalyst bed to measure the temperature profile.

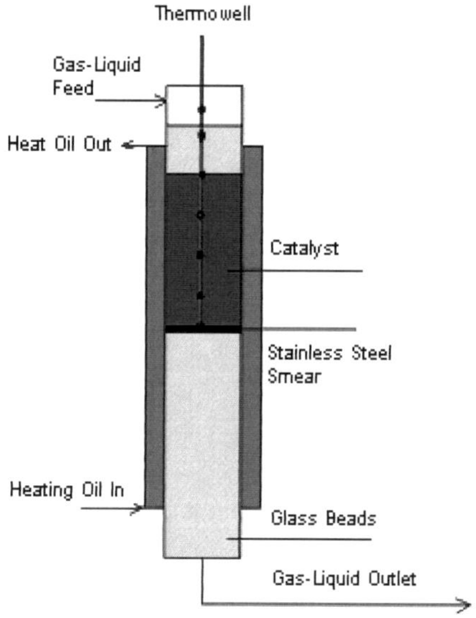

Figure 1. Trickle bed configuration

Experiments were done with 40 wt% glycerol (99.9%, J.T.Baker) solution with sodium hydroxide (98.7%, J.T.Baker) as a base promoter. Pressure, temperature, and feed flow rate were varied to characterize catalyst/reactor performance. Samples were analyzed via high performance liquid chromatography using a BioRad XP-87H column with 0.5 mM H_2SO_4 as the mobile phase; the RI detector was calibrated via a multipoint calibration curve.

Kinetic Model

The establishment of the reactor model begins with development of a kinetic rate expression based on the accepted reaction mechanism for glycerol hydrogenolysis. In simplified form, the mechanism involves 1) dehydrogenation of glycerol to an adsorbed glyceraldehyde analog (*GA·S*); 2) rearrangement and dehydration of *GA* to a second adsorbed intermediate (*I·S*) analogous to pyruvaldehyde, and 3) hydrogenation of the intermediate to propylene glycol (*PG*).

$$GO + S \underset{k_{-1}}{\overset{k_1}{\rightleftarrows}} GA \cdot S + H_2 \quad (1)$$

$$GA \cdot S + OH^- \underset{k_{-2}}{\overset{k_2}{\rightleftarrows}} I \cdot S + OH^- \quad (2)$$

$$I \cdot S + H_2 \xrightarrow{k_3} PG + S \quad (3)$$

The rate expression for each step shown above can be written as follows:

$$r_1 = -\frac{dC_{GO}}{dt} = k_1 C_{GO} \cdot C_S - k_{-1} C_{GA \cdot S} C_{H_2} \quad (4)$$

$$r_2 = -\frac{dC_{GA \cdot S}}{dt} = k_2 C_{GA \cdot S} C_{OH^-} - k_{-2} C_{I \cdot S} C_{OH^-} \quad (5)$$

$$r_3 = -\frac{dC_{I \cdot S}}{dt} = k_3 C_{I \cdot S} C_{H_2} \quad (6)$$

Equating all three rates, invoking the site balance,

$$C_{total} = C_S + C_{GA \cdot S} + C_{I \cdot S} \quad (7)$$

eliminating the concentrations of adsorbed intermediates, and dismissing two terms in the denominator leads to the final expression:

$$r = r_1 = \frac{k_f C_{H_2} C_{GO} C_{OH^-}}{C_{GO} C_{OH^-} + K_H C^2_{H_2}} \quad (8)$$

Thus, glycerol hydrogenolysis reaction kinetics are described in terms of a pre-exponential factor k_0, a reaction activation energy E_a, and an adsorption constant K_H that is independent of temperature. This rate expression captures essential features of glycerol hydrogenolysis: partial order with respect to glycerol and hydroxide, and negative order with respect to hydrogen at high pressures.

Reactor Model

The trickle bed model for glycerol hydrogenolysis is developed for steady state operation by assuming one-dimensional plug flow and using the glycerol molar balance and the reactor energy balance. The molar balance for glycerol is given by

$$-\frac{dC_{GO}}{d\tau} = R_{obv} \quad (9)$$

and the energy balance includes terms for heat generation by reaction and energy removal by the coolant.

$$\frac{d(\rho C_p T)}{d\tau} = (-\Delta H_R)(-R_{obv}) + \frac{2U_0}{R}(T_c - T) \tag{10}$$

Mass Transfer

In order to calculate the hydrogen concentration at the catalyst particle surface ($C_{H2,\ S}$), the concentration gradient for each compound in steady state three phase interface is defined as shown in the Figure 2. Inside the catalyst particle, the observed reaction rate R_{obv} is a summation of reaction rates occurring in both wetted and unwetted factions of the trickle bed. This approach is taken because liquid velocities in the reactor are low under our operating conditions and the bed is only partially wetted [11]. In the wetted fraction, mass transport resistances across the liquid film surrounding the catalyst particles must be accounted for (Figure 2); hence the rate expression must include hydrogen concentration expressed as $C_{H2,S}$. In the unwetted fraction, there is no liquid layer surrounding the catalyst particle and thus hydrogen gas has direct access to the catalyst surface. Its concentration can be taken as the gas phase equivalent, C^*_{H2}. It is assumed that the liquid in both wetted and unwetted fractions is replenished continuously.

$$-R_{obv} = \varepsilon_W(1-\varepsilon_B)\frac{\eta_{H_2} k_f C_{H_2,S} C_{GO} C_{OH^-}}{C_{GO}C_{OH^-} + K_H C^2_{H_2,S}} + (1-\varepsilon_W)(1-\varepsilon_B)\frac{\eta_{H_2} k_f C^*_{H_2} C_{GO} C_{OH^-}}{C_{GO}C_{OH^-} + K_H C^{*2}_{H_2}} \tag{11}$$

The hydrogenolysis reaction is hydrogen limited in the catalyst particles, because of the low solubility of hydrogen in water (~0.05 M) in comparison to the glycerol concentration used (1.0~4.8 M). A first order approximation is therefore used for the effectiveness factor for hydrogen intraparticle diffusion:

$$\eta_{H_2} = \tanh\phi \big/ \phi \tag{12}$$

where

$$\phi = \frac{dp}{6}\sqrt{\frac{k_f C_{GO}}{\varepsilon_p{}^2 D_{H_2}}} \tag{13}$$

At steady state in the wetted fraction of the bed, hydrogen flux across gas-liquid and liquid-solid interface should equal hydrogen consumption rate in the catalyst. Therefore the following relations can be established:

$$k_{GL,H_2}a(C_{H_2}{}^* - C_{H_2,L}) = k_{SL,H_2}a(C_{H_2,L} - C_{H_2,S}) = \eta_{H_2}(1-\varepsilon_B)\frac{k_f C_{H_2,S} C_{GO} C_{OH^-}}{C_{GO}C_{OH^-} + K_H C^2_{H_2,S}} \tag{14}$$

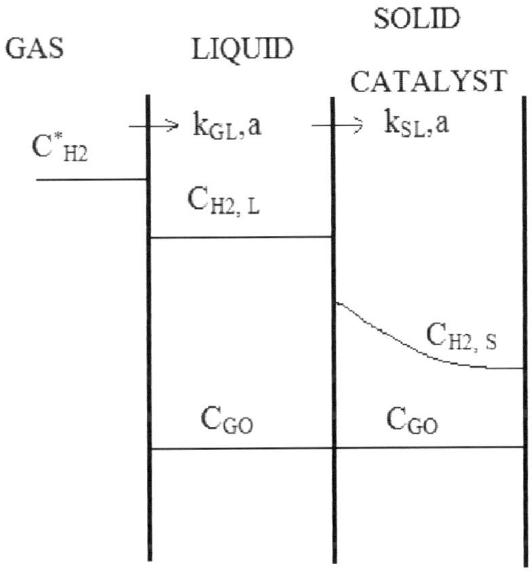

Figure 2. Steady-state concentration profiles in the wetted fraction of the trickle bed reactor.

$C_{H2,\,S}$ can be solved by the following cubic equation which is derived from the equation group with respect to $C_{H2,\,S}$ and $C_{H2,\,L}$:

$$\begin{aligned} &\frac{k_{GL,H_2} a k_{SL,H_2} a k_H}{k_{GL,H_2} a + k_{SL,H_2} a} \cdot C^3_{H_2,S} + \left(\frac{k_H k^2_{GL,H_2} a^2 C^*_{H_2}}{k_{SL,H_2} a + k_{GL,H_2} a} - k_{GL,H_2} a k_H C^*_{H_2} \right) \cdot C^2_{H_2,S} \\ &+ \left(\eta_{H_2} (1-\varepsilon_B) k_f + \frac{k_f k_{GL,H_2} a k_{SL,H_2} a}{k_{GL,H_2} a + k_{SL,H_2} a} \right) \cdot C_{GO} C_{OH^-} C_{H_2,S} \\ &+ \frac{k_f k^2_{GL,H_2} a^2 C^{*\,2}_{H_2} C_{GO} C_{OH^-}}{k_{SL,H_2} a + k_{GL,H_2} a} - k_f k_{GL,H_2} a C^*_{H_2} C_{GO} C_{OH^-} = 0 \end{aligned} \qquad (15)$$

The mass transfer coefficients $k_{GL,H2}$ and $k_{LS,H2}$ were determined for the trickle bed correlations of Larachi et al [6] and Huang et al [8] respectively. The fractional wetting of the trickle bed was found from Al-Dahnam et al [2]. All other physical property parameters were taken from the literature [12] or were estimated from existing correlations; values are given in **Table 1**.

Table 1. Operation parameters used for trickle bed modeling

Description	Unit	Value
Particle size (dp)	m	0.0008
Particle porosity (epsilon)		0.6
Particle wet density (rhop)	kg/m^3	1400
Catalyst Dry Bulk	kg/m^3	400

Density		
Bed depth (Z)	m	0.272947
Bed porosity (epsilonB)		0.5
Reactor radius	m	0.00635
Liquid viscosity (muL)	kg/m/sec	0.00014
Liquid density (rhoL)	kg/m^3	1100
Gas density (rhog)	kg/m^3	1.065302
Gas viscosity (mug)	kg/m/sec	1.38E-05
Glycerol liquid diffusivity	m^2/sec	1.27E-08
Hydrogen liquid diffusivity	m^2/sec	5.5E-08
Liquid surface tension	kg/sec^2	0.0432

Solution Method and Kinetic Parameter Estimation

The molar and energy balance equations were numerically integrated over reactor length using Euler's method in a program written in Microsoft Excel. At each point along the integration, the surface hydrogen concentration $C_{H2,S}$ was determined via the cubic equation shown above. Then all the concentrations and rates were calculated at that point, and the step to the subsequent point was taken.

The reactor model was applied to a large set of steady state trickle bed reactor runs at widely varying conditions using the same catalyst bed. The outlet glycerol conversion was predicted at each set of conditions from the model, and then the kinetic parameters in the rate expression (Eq. 8) k_o, E_a, and K_H and the overall heat transfer coefficient U_o were varied to minimize the sum of square differences between experimental and predicted glycerol concentrations. An initial value of activation energy E_a was calculated based on the data from two experiments running at different temperatures. Because the reaction is only partial order in glycerol, the objective function used in the regression was for a ½ order reaction.

Results

The optimized values of the kinetic parameters and U_o are listed in **Table 2**.

Table 2. Optimal Kinetics Parameters for rate expression (Eq. 8)

Description	**Unit**	**Value**
Preexponential factor k_o	M^3/kg catalyst /sec	9.85E+8
Activation energy E_a	kJ/kmol	86000
Hydrogen coeff k_H	-	425
Heat transfer coeff. U_o	$J/sec/m^2/K$	100

A comparison of experimental and predicted outlet glycerol concentrations over a wide range of operating conditions is shown in Figure 3; it is seen that there is good agreement between predicted and experimental values over all conditions. A sample temperature profile along the bed is given in Figure 4; again, the model predicts with good accuracy the exotherm of about 10 K in the trickle bed reactor.

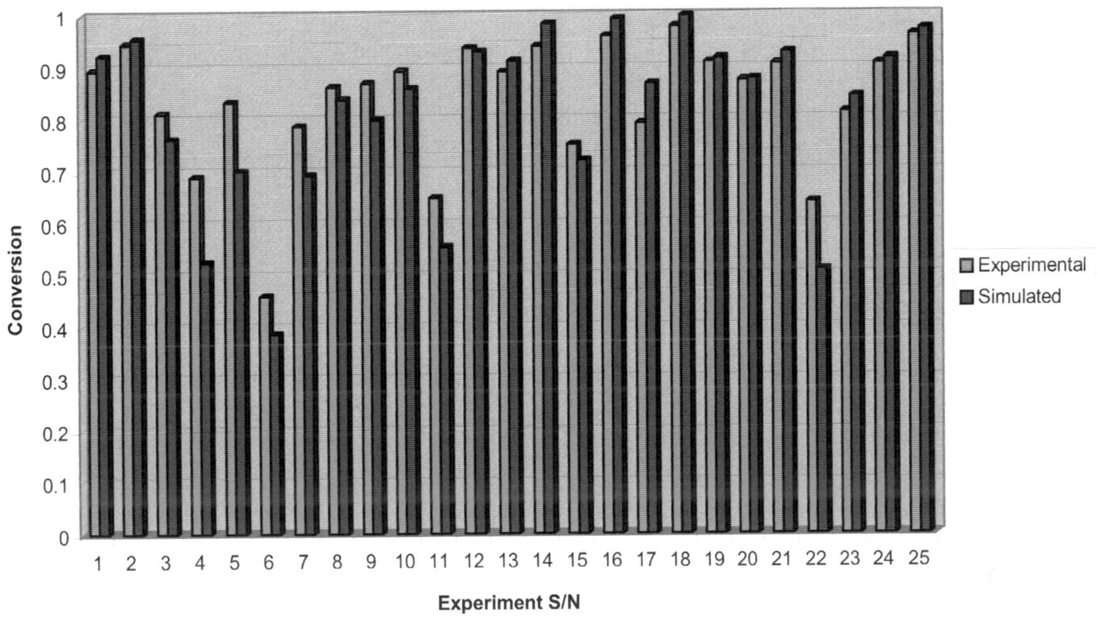

Figure 3. Outlet glycerol simulated and experimental concentrations

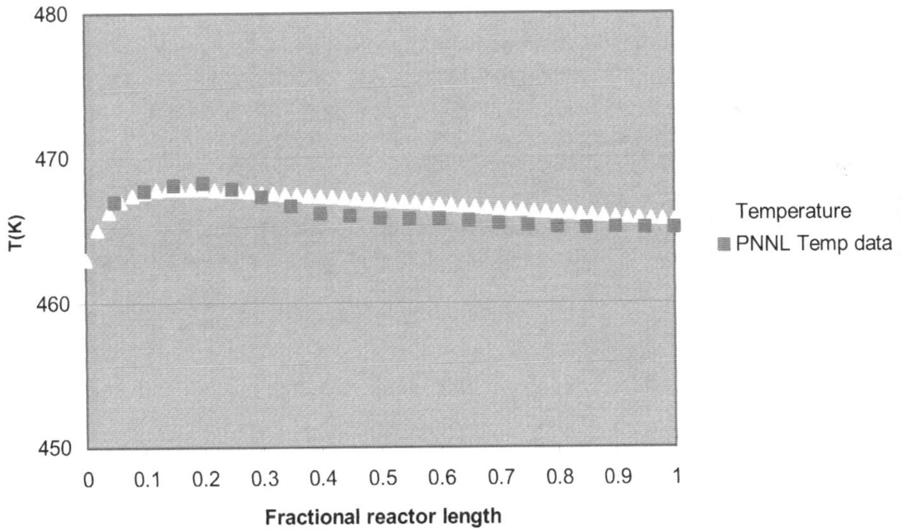

Figure 4. Predicted vs. experimental temperature profile in trickle bed (Exp. 7)

Based on our current work, an initial commercial scale up applying our model was done with yearly capacity of 100 MM lb. Outlet glycerol conversion is over 90% with the right trickle bed configuration and operating conditions.

Reference

1. Dudukovic, Milorad P. and Larachi, Faical et al (1999), "Multiphase reactors-revisited," *Chem Engi Scie,* pp 1975-1995.
2. Al-Dahhan, Muthanna H. and Larachi, Faical et al(1997), "High-pressure tickle bed reactors: A
   review," *Ind. Eng. Chem. Res.* 36, 3292-3314.
3. Dudukovic, M.P(1999), "Trends in catalytic reaction engineering," *Catalysis Today.* Pp 485-15.
4. Biardi, Giuseppe, Baldi, Ciancarlo(1999), "Three-phase catalytic reactors." *Catalysis today.* 52 pp 223-234.
5. Dianetto A, Specchia, V(1992), "Trickle-bed reactors: state of art and perspectives," *Chemi Engi Sciences.*, 47 pp 3197-3213.
6. Larachi, F. Cassanello M et al(1998), "Gas-liquid interfacial mass transfer in trickle-bed reactors at Elevated Pressures," *Ind. Eng. Chem. Res.*, 37, pp718-733.
7. Iliuta, Ion, Ortiz-Arroyo and Arturo et al (1999), "Hydrodynamics and mass transfer in trickle bed reactor: an overview," *Chem Engi Scien.,* 54 pp5329-5337.
8. Huang, X. and Varma,A. et al(2004), "Heat transfer characterization of gas-liquid flows in a trickle bed," *Chem. Engi. Sciences.* 59, pp 3767-3776.
9. Al-Dahhan Muthanna H. and Dudukovic,Milorad P(1995), "Catalyst Wetting Efficiency in trickle bed reactors at high pressure," *Chem Engi Sci.,* 50 pp2377-2389.
10. Tronconi E.,Ferlazzo, N. et al(1992), "A mathematical model for the catalytic hydrogenolysis of carbohydrates," *Chemical engi. Sci.,* 47, pp 2451-2456.
11. Dietz, A., Julcour, C. et al. (2003), **"**Selective hydrogenation in trickle-bed reactor: experimental and modelling including partial wetting,**"** *Cata. Today*, 79, pp 293-305.
12. Perry, R.H., Green, D.W. and Maloney, J.O.(1997), "Perry's Chemical Engineering Handbooks," New York, McGraw-Hill.

404c Potential For Producing Propylene Oxide And Other Oxygenated C3 Hydrocarbons From Propylene Glycol Derived From Glycerin

Surupa D. Abraham and Eric J. Doskocil. University of Missouri - Columbia, W2033 Lafferre Hall, Dept. of Chemical Engineering, Columbia, MO 65211

Propylene oxide is currently produced commercially from propylene isolated as a co-product of petroleum processing. The reaction processes are based on chlorohydrin or hydroperoxide synthesis procedures, which are very hazardous. These processes are inefficient, energy intensive, involve multiple steps, have low selectivity, and are accompanied by huge volumes of effluent wastewater and by-products. In addition, direct oxidation catalysts are very expensive due to the fact that they contain precious metals, like gold and silver. It is evident that propylene oxide production could be significantly enhanced by finding an alternative production route than starting with a propylene feedstock.

When soybean oil is used to produce biodiesel, significant quantities of glycerin are produced as a co-product. Recently, commercial production of propylene glycol from glycerin has begun. This propylene glycol has the potential to be a feedstock that can be further reacted to produce propylene oxide and other oxygenated C3 hydrocarbons that are commonly produced using a propylene feedstock, providing a renewable route to common petroleum-based C3 hydrocarbons.

The limited literature covering the production of propylene oxide from propylene glycol indicates that basic catalysts are optimum for this reaction to occur selectively to propylene oxide. Our studies have focused on using common base catalysts, including magnesium oxide, calcium oxide, and alkali modified aluminas and zeolites. We have identified nine total products that have been produced from our gas-phase reactivity studies: acetaldehyde, propylene oxide, propionaldehyde, acetone, 2-propanol, 1-propanol, allyl alcohol, acetol, and dipropylene glycol. The effect of catalyst type, temperature, and space time on selectivity to propylene oxide will be discussed and potential reaction pathways explored. We have found that, when the residence time of propylene glycol in the reactor is decreased, an increase in selectivity to propylene oxide is observed, indicating that propylene oxide may be a reactive intermediate in the production of the other byproducts formed. In addition, the production of dipropylene glycol appears to contribute to higher propylene oxide production rates. While the primary goal of this research was to produce propylene oxide, catalysts that are effective at producing any other of the other byproducts at high selectivity may be useful routes for the renewable production of other high-volume bulk chemicals that are typically produced from propylene.

404d Biofuel Production By Catalytic Fast Pyrolysis

Tushar Vispute, University of Massachusetts - Amherst, Amherst, MA 01003, Torren Carlson, Chemical Engineering, University of Massachusetts - Amherst, Amherst, MA 01003, and **George Huber**, Chemical Engineering, University of Massachusetts-Amherst, 159 Goessmann Lab, 686 North Pleasant St., Amherst, MA 01003.

Concerns about national security, global warming, combined with the diminishing supply and increased cost of fossil fuels are causing our society to search for new sources of transportation fuels. In this respect the only sustainable source of renewable carbon is plant biomass. Cellulosic biomass is both inexpensive ($15 per barrel of oil energy equivalent) and abundant (the US can sustainably produce 4 billion barrels of oil energy equivalent per year). Fast pyrolysis is one of the most promising methods for biofuel production. Fast pyrolysis involves rapidly heating the biomass (500oC/sec) to intermediate temperatures (400-600oC) followed by rapid cooling (residence times 1-2 s). These reaction conditions allow the conversion of thermally unstable biomass compounds to a liquid product called bio-oils while minimizing undesired reactions (i.e. coke and gas formation). The bio-oils are an acidic combustible liquid containing more than 300 compounds that degrade with time. It would be desirable to improve the characteristics of the bio-oils produced in the fast pyrolysis process. As we will show in this presentation addition of heterogeneous catalysis to fast pyrolysis (catalytic fast pyrolysis) significantly changes the products produced. The fast heating rates used in fast pyrolysis allow the conversion of thermally unstable compounds allows us to convert these compounds before they produce undesired coke.

We have tested a wide range of catalysts (including SiO2-Al2O3, Pt/ SiO2-Al2O3, WOx/ZrO2, SOx/ ZrO2, HY-zeolite, â-zeolite, and HZSM-5) for feedstock-derived from cellulosic (xylose, glucose, xylitol, cellobiose, cellulose) and lignin (organosolv and benzyl phenyl ether) fractions of biomass in a microscale flash pyrolysis reactor. The reactions involved for catalytic fast pyrolysis include dehydration, hydrogen transfer, decarbonylation and aromatic production. These reactions must be properly balanced to produce the desired reaction. Shape selectivity appears to be very important for these reactions as zeolite catalysts produce more aromatics than amorphous catalysts.

We have used xylitol as a feed to study the effects of different catalysts. Coke, dehydrated products (including furan, methyl furan and furfural), CO and CO2 were the primary products for SiO2-Al2O3, Pt/ SiO2-Al2O3, WOx/ZrO2, and SOx/ZrO2. Coke, aromatics and CO were the primary products for HZSM-5. The products for HY-Zeolite and â-Zeolite include coke, dehydrated products, aromatics, CO and CO2. The aromatics produced in order of decreasing concentration include: alkyl substituted Naphthalenes >> Toluene > Xylenes > Naphthalene > Benzene > alkyl substituted Indanes >> Indane. These aromatics could be used as a gasoline blend. Other cellulosic feeds (including cellulose, cellobiose-glucose dimer, and glucose) gave similar product selectivities with a â-Zeolite catalyst suggesting that catalytic fast pyrolysis can be used for conversion of polysaccharides to aromatics.

We also tested catalytic fast pyrolysis of organosolv lignin and benzyl phenyl ether (BPE-a model lignin compound). Pyrolysis of BPE produced phenol, benzene and toluene as the major products. Pyrolysis of neat BPE resulted in phenol (80% selectivity) as the main product whereas in presence of Silica-Alumina or HZSM-5 as catalyst phenol and benzene were the major products with about 40% selectivity The selectivity changes from phenol to benzene as the catalyst to BPE ratio increases. Catalytic fast pyrolysis of organsolv lignin with HZSM-5 and â-zeolite produced aromatics (40 % selectivity) and CO (45 % selectivity).

In summary heterogeneous catalysts can significantly change the reaction chemistry that is occurring in the fast pyrolysis process. It is likely that future advances in understanding the reaction mechanism combined with new material synthesis and theoretical chemistry will help improve the reaction chemistry further and make catalytic fast pyrolysis an efficient reaction for biofuel production.

EFFECT OF METALLIC COMPOUNDS ADDITION ON THE DECREASE IN TAR YIELD DURING PYROLYSIS OF BIOMASS

Yoshikage Ohmukai[a], Kosuke Fujimoto[a], Isao Hasegawa[a], Shigeya Hayashi[b], Kazuhiro Mae[a]
[a]Kyoto University, Kyoto, Japan, [b]Ube Industries, Ltd., Ube, Japan

Introduction

Biomass is one of attractive feedstocks for realizing carbon neutral. Now, various conversion technologies have been developed to utilize biomass efficiently. In these technologies, pyrolysis is promising one because of its simplicity. However, a large amount of tar evolves through pyrolysis and causes a trouble in the process. Therefore, an effective method for suppressing tar formation is desired. The strategy to suppress tar formation is classified into two approaches: One is a post treatment such as catalytic steam reforming after pyrolysis. The other is a pretreatment of biomass structure. Biomass consists of cellulose, hemicellulose, and lignin. Cellulose component is mainly decomposed into tar from 350 to 500 °C, so it is considered that the modification of cellulose structure is essential to diminish tar formation. Raveendran et al. [1] and Fahmi at al. [2] examined the influence of mineral matter on the biomass pyrolysis. They suggested that the inorganic species in biomass strongly affected on the biomass pyrolysis behavior. This suggests that the addition of metallic compound may change the cellulose structure by the interaction between metal and the functional groups in biomass. In this study, we added several kinds of inorganic metallic compounds to biomass in order to examine the effect on the suppression of tar formation.

Experimental

Materials

The raw materials used in this study are Douglas fir and paper cellulose. Table 1 lists the ultimate analyses of the samples. All materials were pulverized under 500 μm and dried *in vacuo* for 24 h at 70 °C prior to use. Metal hydroxides powders were simply mixed with the biomass samples with appropriate amount of pure water. The mixed sample was dried again *in vacuo* for 24 h at 70 °C. The amount of inorganic ion loaded on samples per mixed sample was fixed at 8.3 wt%.

Pyrolysis of biomass

The samples were pyrolyzed using a thermogravimetric analyzer (TGA-50;

Table 1 Ultimate analysis of feedstocks

Sample	Ultimate analysis (wt%daf)		
	C	H	O(diff.)
Douglas fir	47.8	5.5	46.7
cellulose	42.9	5.9	51.2

Shimadzu Co. Ltd.) at a heating rate of 20 K min^{-1} from room temperatures to 110 °C in a nitrogen flow of 0.83 ml (STP) s^{-1}. After keeping at 110 °C for 30 min, the sample was heated up to 600 °C at a rate of 20 K min^{-1}

Apart from the TGA experiment, the pyrolysis using a tubular reactor was conducted to prepare the char samples for analysis. Fig. 1 shows the schematic of experimental apparatus. The quartz reactor tube of 10.7 mm i.d. was used. Samples were loaded on a ceramic boat, and the boat was set at the center of the heating zone. Nitrogen gas was fed at a rate of 0.83 ml (STP) s^{-1}. 50 mg of samples were heated to a final temperature of 500 °C at 20 K min^{-1} and were kept at the final temperature for 10 min. Gaseous products obtained through pyrolysis were collected in a gas bag, and their yields were determined using a gas chromatograph (Shimadzu, GC-14A) equipped with Porapak TypeQ column or Molecular Sieve 5A. The experiment was conducted at least three times under the same conditions to ensure the reproducibility of data.

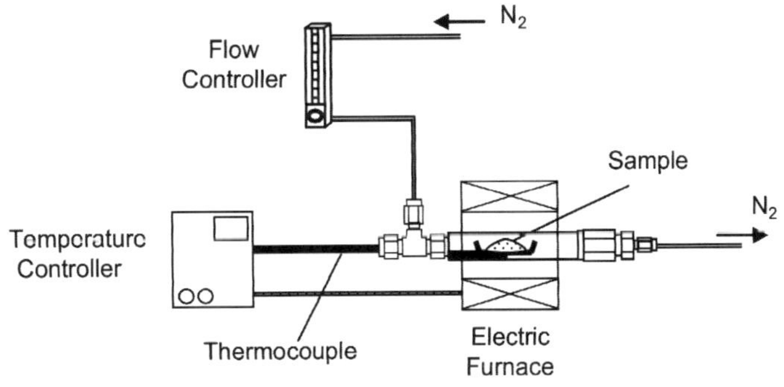

Fig. 1. Schematic diagram of the reactor used in pyrolysis

Characterization of char

The elemental composition of the char is determined from the ultimate analysis using a CHNS-O analyzer (ECS4010; Costech Instruments). The change in the functional groups in biomass char was analyzed using an FT-IR spectrometer (JIL-SPX60; JEOL Ltd.).

Results and discussion

Effects of metal hydroxides on pyrolysis of biomass

First, we examined the effect of metal hydroxide loaded on the pyrolysis profile of Douglas fir. The metals used were Na, K, Mg, Ca, Ba, Fe, Cu, Zn, and Al. Among them, barium hydroxide forms is an octahydrate, zinc hydroxide form is $ZnO{\cdot}H_2O$, and iron hydroxide form is FeO (OH). The weight change profiles during the pyrolysis of metal-loaded samples are shown in Fig. 2-(a). The metal addition affected weight change profile, namely, the temperature region of main decomposition of biomass and the char yields. The profiles were classified in two groups: One is the group whose profile shifted to low temperatures (group I; Na, K, Ba), and the other is the group whose profile shifted to high temperature (group II; Ca, Mg). This difference in the profile suggests the pyrolysis mechanisms are different between two groups. For the group I, the reaction between 200 to 300 °C was promoted, suggesting that inter- and intramolecular cross-linking of cellulose occurred by dehydration. For the second group, on the contrary, the reaction was delayed to high temperature, suggesting that the decomposition of functional groups in biomass was suppressed with keeping the structure of raw state. The suppression of decomposition is confirmed later by examining the amount of functional groups in the char.

By contrast, addition of hydroxides of Al, Fe, and Zn into biomass did not influence on the pyrolysis profile at all. Hasegawa [3] has shown the effects of metal hydroxides on pyrolysis of wet coffee wastes. Adding calcium hydroxide to the wastes increases char yield while copper hydroxide has no effects on the pyrolysis behavior. These results show that the effect of metal hydroxide addition to biomass mainly depends on the species of metal compounds rather than that of biomass.

The increase in char yield by the hydroxides addition was in the order

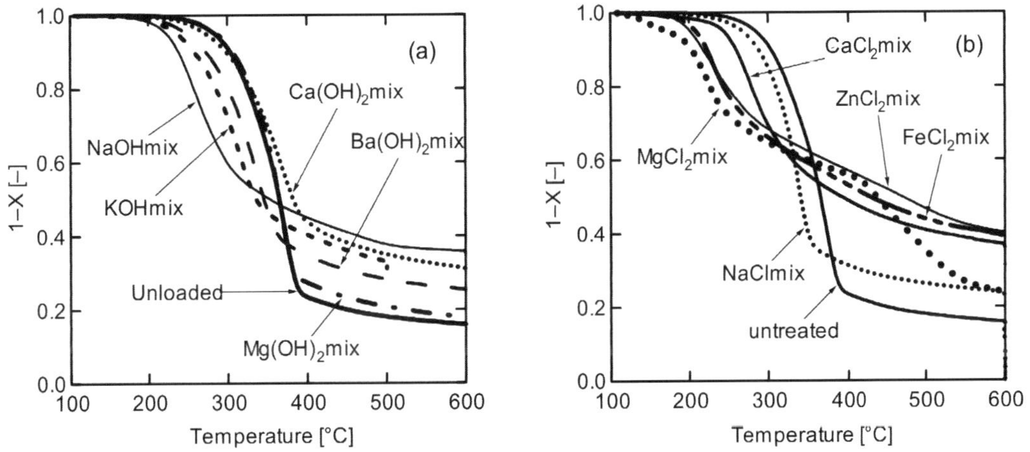

Fig. 2. The pyrolysis profiles of Douglas fir with various metal additives.

Na>Ca>K>Ba>Mg. The hydroxides brought about the increase in char yield have common characteristics of high ionicity and high solubility in water.

Effects of metal chlorides on pyrolysis of biomass

Metal chlorides also had high solubility in water, a similar effect is expected. Pyrolysis profiles of Douglas fir with chloride compounds of Ca, Mg, Fe, Zn, or Na were measured using TGA as shown in Fig. 2-(b). The weight of ash remained after pyrolysis and combustion was subtracted from each TG curve. The figure shows that weight change took place at around 250 °C in pyrolyses of all the chloride-loaded samples. Moreover, a characteristic kinked profile in TG curve of Douglas fir at 390 °C did not appear when the samples loaded metal chloride compounds were pyrolyzed. These results clearly show that metal chlorides have an effect on the pyrolysis behavior of biomass as the category of group I. Namely, metal chloride has high affinity for water such as solubility and deliquescency. The strong hydrophilicity may cause the subtraction of water from biomass during pyrolysis and promote cross-linking of cellulose. Another effect of metal chloride is high solubility by water. This brings about a uniform loading by a molocular level.

Tar suppression effect

Fig. 3 shows the product yields by the pyrolyses of Douglas fir with hydroxides. The yields represents on the basis of water and ash free. The amount of ash in the residue of Na-loaded biomass was determined after combustion of char excluding Na which was estimated as a form of NaOH. The char yields of untreated biomass, Ca-loaded sample, and Na-loaded were 19, 29, and 32 wt%, respectively. Addition of metal hydroxides to biomass caused a significant change on the product distribution. Gas yield of untreated biomass in pyrolysis was ca. 9%, whereas those of Ca-loaded and Na-loaded biomass were 16% and 26%, respectively. Especially, tar yield of Na-loaded biomass in pyrolysis was nearly zero while that of untreated

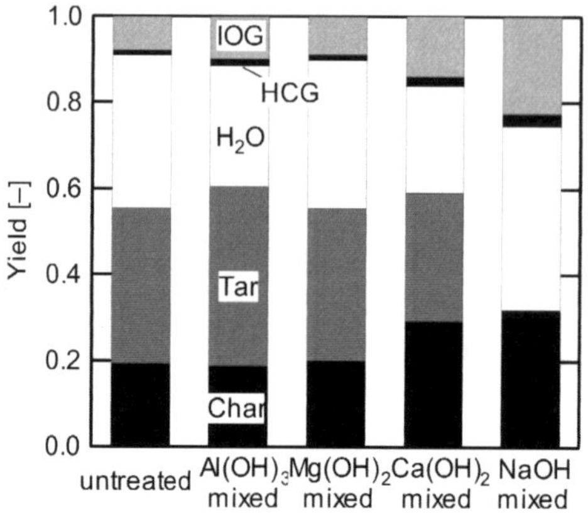

Fig. 3 Product distributions in pyrolyses of hydroxides-loaded Douglas fir at 500 °C.

biomass was ca. 40%. The adhesion of tar to inner wall of the reactor was not observed during the pyrolysis of Na-loaded sample. These results show that sodium hydroxide remarkably suppresses tar evolution during pyrolysis of biomass and gives char in high yield.

Estimation of pyrolysis mechanism

NaOH may be one of the effective additives for suppressing the tar formation and increasing the char yield during pyrolysis. To clarify the reason, the difference in pyrolysis behavior with and without NaOH was examined. Fig. 4 compares the changes in the i.r. spectra of O-H stretching vibration during pyrolysis between untreated biomass and NaOH-loaded one. For the untreated biomass, the crystal structure of cellulose is slightly distorted at lower temperatures and partly cross-linked. However, most part of cellulose is volatilized as tar through the cleavage of glycosidic linkage and the formation of cellulose monomer or dimmer at around 350 °C. By contrast, for the NaOH-loaded sample, the interaction of Na^+ ion and hydroxyl groups distorts the crystal structure at lower temperatures. As the distortion shortens the distance between hydroxyl groups on cellulose, cross-linking is formed at low temperature via dehydration reaction. This is supported by the result that the water yield is extremely high for Na-loaded samples. Such a highly cross-linked structure formed at low temperature was kept as a solid state when the cleavage of glycosidic linkage partly occurred above 350 °C. This brought about the significant suppression of the tar evolution.

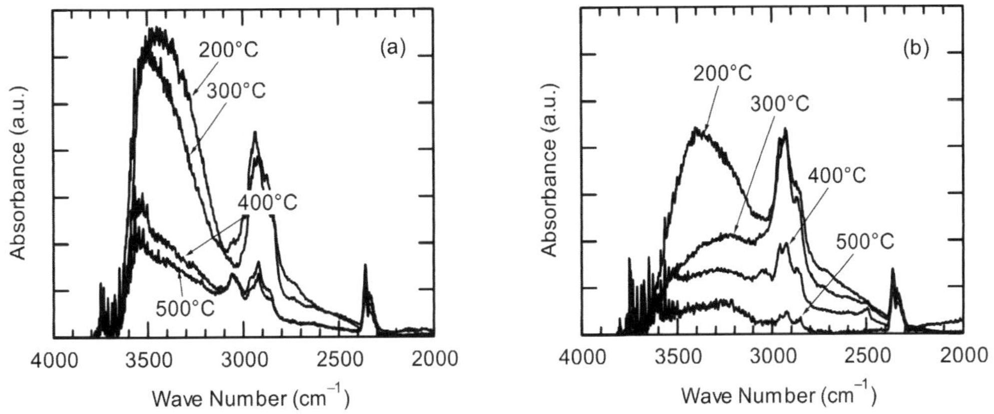

Fig. 4. FT-IR spectra of char prepared through pyrolysis at 200–500 °C: (a) unloaded biomass; (b) NaOH-loaded.

Effect of the amount of NaOH loaded on biomass

In order to examine quantitatively the effect of NaOH on pyrolysis yield, we conducted the pyrolysis of paper cellulose with various amounts of NaOH. The weight ratio of Na^{+} ion to the cellulose was ranged from 0.5 to 8.3 %. Fig. 5

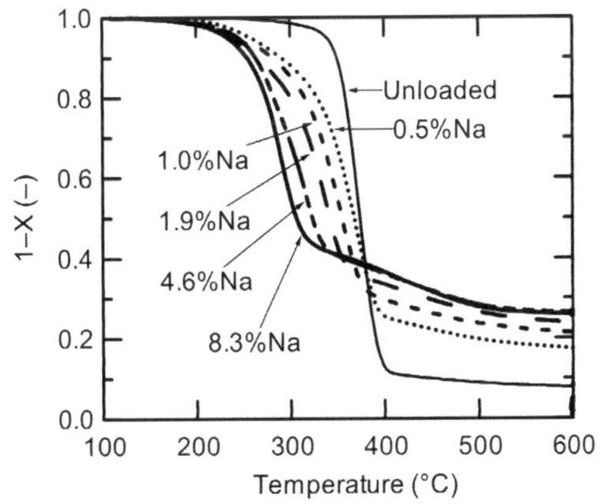

Fig. 5. The effect of the amount of sodium hydroxide added to cellulose on pyrolysis profile.

shows the change in the pyrolysis profile by adding NaOH. The char yield at 500 °C increased from 9% to 20% even by adding 0.5% of Na. With the increase in the amount of Na from 0.5 to 4.6 %, the main decomposition region shifted to low temperature and the char yield at 500 °C increased. Comparing the effect of the amount of Na on the char yield in Fig. 5, it was confirmed that 4.6 % addition of Na was sufficient to suppress the devolatilization.

As mentioned above, it is presumed that the pyrolyzed intermediate structure of biomass at lower temperatures is quite different in the presence of Na. Then, we introduced a degree of cross-linking to indicate how far cross-linking of the intermediate proceeds. The degree of cross-linking, *C*, was defined as a fraction of hydroxyl groups that was consumed in dehydration reaction and produced water at lower temperatures to all the hydroxyl groups of cellulose. Accordingly, *C* is given by

$$C = \frac{\text{The number of cross-linking}}{\text{The number of hydroxyl groups in cellulose before pyrolysis} \times 0.5} \quad (1)$$

the number of hydroxyl groups is multiplied by 0.5 because two hydroxyl groups make a molecule of water. The amount of water produced from the intermediate during pyrolysis is here substituted for the number of cross-linking in cellulose as a first approximation. Cellulose was pyrolyzed for 1 h at 200 °C using TGA so that further decomposition such as tar evolution should not proceed at higher temperatures. The weight decrease during pyrolysis at 200 °C was assumed to be derived from only water formation. Each cellulose monomer has three hydroxyl groups, and the molecular weight of the monomer is 162 g mol^{-1}, i.e., one and a half molecule of water is formed per the monomer when all the hydroxyl groups are consumed for cross-linking reaction. The complete cross-linking corresponds to 16.7 wt% of weight reduction. We substitute the weight reduction at 200 °C in the eq. (1) to obtain

$$C = \frac{\text{The weight reduction of the intermediate}}{16.7\ \text{wt\%}} \quad (2)$$

The TGA profiles shown in Fig. 5 suggest that the amount of dehydration at lower temperatures is closely related with each char yield at higher temperatures. Then, the relationship between the degree of cross-linking and the char yield at 500 °C was examined. Fig. 6 shows the relationship between the char yield at 500 °C and the index, C. The char yield correlated in linear with the degree of cross-linking. This result indicates that the increase in char yield is controlled by dehydration reaction at lower temperatures. Furthermore, the linear relationship suggests that the increase in the amount of cross-linking directly affects the char yield rather than that mediated through some steps.

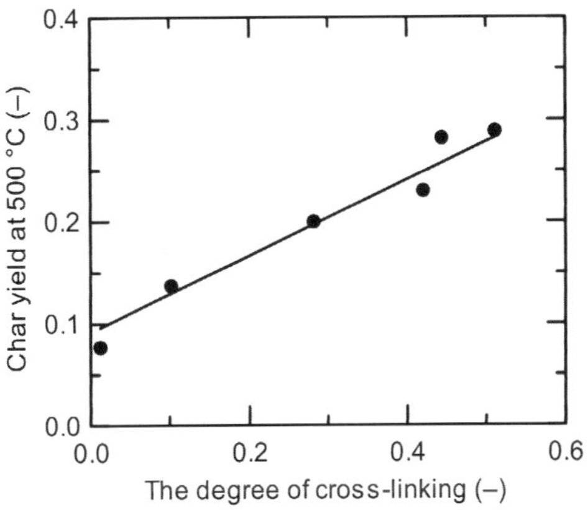

Fig. 6 The relationship between the degree of cross-linking and char yield at 500 °C during pyrolysis.

Conclusion

The effect of metal compounds on biomass pyrolysis was examined, then it was found that sodium hydroxide has a significant effect for the suppression of the tar evolution and the increase in char yield. Main decomposition region significantly shifted to lower temperatures by the addition of Na. It was clarified that the drastic change in pyrolysis behavior was caused by the distortion of the crystal structure of cellulose and the promotion of cross-linking at lower temperatures. Based on these results, we introduce the index for the degree of cross-linking. The char yield at 500 °C correlated in linear with the degree of cross-linking. From these results, we can say that the pyrolysis of biomass can be controlled precisely through understanding of the change in structure of the intermediate.

Reference

1. Raveendran, K., Anuradda Ganesh and Kartic C. Khilart (1995), "Influence of mineral matter on biomass pyrolysis characteristics," *Fuel*, **74**, pp. 1812–1822.
2. Fahmi, R., A. V. Bridgwater, L. I. Darvell, J. M. Jones, N. Yates, S. Thain and I. S. Donnison (2007), "The effect of alkali metals on combustion and pyrolysis of *Lolium* and *Festuca* grasses," *Fuel*, **86**, pp. 1560–1569.
3. Hasegawa I., S. Horiguchi and K. Mae (2006), "Suppression of tar formation by co-pyrolysis of calcium-loaded biomass and low temperature bio-oil," in 232nd American Chemical Society National Meeting & Exposition.

404f Process Modeling And Economic Evaluation Of Mix Alco Process To Produce Various C2 To C7 Chemicals

Parag S. Shah and Peter van Walsum. Environmental Studies, Baylor University, Goeble Building, Waco, TX 76710

Mixalco process is one of the several bioprocesses developed in past few years to convert different cellulosic biomass to value added products. The mix alco process can be designed to produce chemicals such as organic acids, ketones or alcohols as final products. In this project the process modeling and economic evaluation of the mix alco process to produce organic acids, ketones and alcohols was performed. The fixed capital investment, operating cost and return on investment of the process to produce mixed alcohols, ketones and volatile fatty acids from different biomass were evaluated and compared. The Superpro Designer 7.0 from Intelligen Inc. was used to do process simulation and design. Manure and sewage sludge were used as principal raw materials.

Process models were prepared to compare the economic aspects of a facility to process 40 tons/hr of manure to alcohols, ketones or lower fatty acids as final products. Sensitivity analysis on key parameters affecting the process was also performed. Finally economic comparisons of the mix alco process with normal anaerobic digestion process to produce biogas for heat and electricity were made.

471a High-Throughput Chemical Analysis of Switchgrass Feedstocks

Bonnie R. Hames, Tanya M. Kruse, Thomas W. Johnson, and Amr Ragab. Ceres, Inc., 1535 Ranch Conejo Blvd., Thousand Oaks, CA 91320

At Ceres, we are applying high-throughput methods for measuring the chemical composition of cellulosic feedstocks as a key element in the development of dedicated bioenergy crops with enhanced characteristics for biofuels production such as high ethanol yields and low cost of conversion. These new techniques combine Near InfraRed (NIR) spectroscopy and Projection to Latent Structures (PLS) multivariate analysis in methods inexpensive enough to allow the compositional analysis of hundreds of samples in one day with precision and accuracy that closely matches wet chemical methods. The development and demonstrated use of rapid analysis methods for chemical characterization of switchgrass feedstocks will be described. The ability to accurately sample a bulk feedstock and to analyze hundreds of samples provides a new tool that is being used to assess the compositional variability of switchgrass as a function of variety, geographical location, agronomics, harvesting method and storage. Studies of this type require thousands of samples and would be too costly to pursue without the savings in time and cost provided by NIR/PLS rapid analysis methods. Furthermore, these analytical methods report data in a form that can be integrated directly into engineering, economic, and life-cycle models for a clearer evaluation of enhanced value across a wide variety of biomass conversion processes.

471b Lignocellulosic Biomass Quality Changes in Relation to Available Feedstock Tonnage

Corey Radtke[1], Douglas L. Karlen[2], Stuart Birrell[3], Lynn Petzke[1], Peter A. Pryfogle[1], and J. Richard Hess[1]. (1) Biotechnology, Idaho National Laboratory, P.O. Box 1625, Idaho Falls, ID 83415, (2) USDA-ARS National Soil Tilth Laboratory, 2150 Pammel Dr., Ames, IA 50011-3120, (3) Department of Agricultural and Biosystems Engineering, Iowa State University, 200C Davidson, Ames, IA 50011

The compositional and functional quality of lignocellulosic biomass impacts their utilization as feedstocks for ethanol production. However, these compositional and functional qualities are in turn skewed by the assembly system used for the biomass feedstock retrieval and delivery. Accessing different levels of biomass tonnage may require different harvesting, storing, and preprocessing operations, which can impact feedstock quality. For example, the mean and variance of corn stover composition is in part a function of the harvested fractions. Concepts and relationships of accessed biomass tonnages and the associated biomass quality will be presented.

471c Influence of Collection Method on the Sugar and Ethanol Processing Characteristics

Michael D. Montross[1], Corey Radtke[2], Czarena Crofcheck[3], and S. A. Shearer[3]. (1) Biosystems and Agricultural Engineering Department, University of Kentucky, 128 Barnhart Building, Lexington, KY 40546-0276, (2) Biotechnology, Idaho National Laboratory, P.O. Box 1625, Idaho Falls, ID 83415, (3) Biosystems & Agricultural Engineering, University of Kentucky, Lexington, KY 40546

Due to concerns with biomass collection systems and soil sustainability there are opportunities to investigate the optimal plant fractions to collect. An ideal feedstock would require a low severity pretreatment to release a maximum amount of sugar during enzymatic hydrolysis that would be available for fermentation. Corn stover fractions were separated by hand and analyzed for glucan, xylan, acid soluble lignin, acid insoluble lignin, and ash composition. The samples were not extracted prior to compositional analysis to most closely match the conditions of a feedstock being delivered from a farm to a biorefinery. The stover fractions were also pretreated with either 0, 0.4, or 0.8% NaOH for 2 hours at room temperature, washed, autoclaved and saccharified. In addition, acid pretreated samples underwent simultaneous saccharification and fermentation (SSF) to ethanol. In general, the two pretreatments produced similar trends with cobs, husks, and leaves responding best to the pretreatments, the tops of stalks responding slightly less, and the bottom of the stalks responding the least. For example, corn husks pretreated with 0.8% NaOH released over 90% (standard error of 3.8%) of the available glucan, while only 45% (standard error of 1.1%) of the glucan was produced from identically treated stalk bottoms. Estimates of the theoretical ethanol yield using acid pretreatment followed by SSF corresponded to 63 (standard error of 11.5%) for husks and 37% (standard error of 1.2%) for stalk bottoms. This suggests that integration of biomass collection systems to remove sustainable feedstocks could be integrated with the processes within a biorefinery to minimize overall ethanol production costs.

471d Strategies to Conduct Effective Co-Fermentation on Cellulosic Hydrolysates from Afex-Pretreated Feedstocks

Ming Woei Lau, Chemical Engineering and Material Science, Michigan State University, 3246 Engineering Building, East Lansing, MI 48824 and **Bruce E. Dale**, Chemical Engeering and Material Science, Michigan State University, 3254, Engineering Building, East Lansing, MI 48824.

Effective utilization of all sugars in cellulosic hydrolysate during fermentation is essential in keeping cellulosic feedstocks a competitive source for various bioconversions including ethanol production. Saccharomyces cerevisiae has long been recognized as a promising microorganism for ethanol fermentation. Recent successes in metabolic engineering enable the strain to utilize xylose as a carbon source, thus allowing co-fermentation of glucose and xylose in cellulosic hydrolysate. However, xylose utilization was found to be bottleneck in ethanol production through co-fermentation of corn-stover hydrolysate at low starting cell density presumably due to inhibitors present in cellulosic hydrolysate.

Hence, we are evaluating the effects of fermentation inhibitors on xylose utilization in co-fermentation. As for possible solutions to increase xylose utilization, the studies on how nutrient supplementation and starting cell density affect xylose utilization are conducted on laboratory media. In addition, similar strategies would be tested in cellulosic hydrolysate derived from AFEX-treated corn stover, DDGS and switchgrass at high solids loading saccharification. These studies are conducted using Saccharomyces cerevisiae 424A(LNH-ST), a genetically-engineered ethanologen for co-fermentation of glucose and xylose.

Web Page: www.everythingbiomass.org

471e Large Scale Biomass Plant in Kansas: Current Status of Biomass

Thomas W. Robb and Jeremy Mall. Abengoa Bioenergy Corporation, 1400 Elbridge Payne Road, Suite 100, Chesterfield, MO 63017

Placeholder Abstract (to be edited):

In North America over 80% of distillers grains are used in ruminant diets. Historically, less than 1% of total annual production is fed to swine. Recent studies have shown however that the nutrition value of distillers grains is becoming more attractive to swine producers. The dry milling industry produces distillers grains, distillers grains plus solubles, and distillers solubles. When fed at 15% or less of the diet dry matter, distillers grains serve as a source of supplemental protein. When fed at higher levels (greater than 15% of the diet dry matter) the primary role is as a source of energy.

498a A Free Energy Landscape Study Of The Reverse Hydrolysis Reactions Of Xylose In Acidic Aqueous Solutions

Haitao Dong[1], Mark Nimlos[2], and Xianghong Qian[1]. (1) Mechanical Engineering, Colorado State University, Mechanical Engineering, 1374 Campus Delivery, Fort Collins, CO 80523, (2) National Renewable Energy Laboratory, 1617 Cole Boulevard, Golden, CO 80401

Biofuels converted from lignocellulosic biomass provide an alternative to the traditional fossil fuels. Pretreatment is one of the crucial steps during biomass conversion to hydrolyze hemicelluloses and relocate lignin. During dilute acid pretreatment, hemicelluloses such as xylan are hydrolyzed into monomer xylose. Under acidic environment and relative high monomer concentration, reverse hydrolysis of xylose can occur thus lowering the sugar yield and the overall biomass conversion efficacy. Here we present a free energy study of the acid-catalyzed reverse hydrolysis of xylose into xylobiose by using the metadynamics sampling technique in the frame of Car-Parrinello molecular dynamics. This history-dependent method allows an efficient sampling of the configurational space along the reaction coordinates and an accurate estimation of the free energy landscape. The reaction pathways and transition states of forming the glycosidic bonds were located on the landscape. The activation energies and free energy changes were also obtained and show a good agreement with experiment. Our estimations of ΔG and understanding of the reaction mechanisms can be used to develop accurate kinetic models for this reverse hydrolysis process.

498b Design of Hydrothermal Treatment Process to Convert Agriculture by-Products into Biofuel and Biochemicals

Kazuhiro Mochidzuki, Univ. of Tokyo, 1-8 Yayoi-cho, Inage-ku, Chiba, 263-0022, Japan and Akiyoshi Sakoda, Institute of Industrial Science, University of Tokyo, 4-6-1 Komaba, Meguro-ku, Tokyo, 153-8505, Japan.

Rice husk, which is a typical agricultural by-product, was treated by a biomass utilization process to obtain biofuel, such as bioethanol, and biochemicals, such as furfural. Steam-explosion treatment was employed as the first step of the process. By this treatment, deconformation of the structure of rice husk and hydrolysis of hemicellulose in the rice husk occurred, simultaneously. The exploded solid product was used as the feedstock of bioethanol after separating water-soluble products. The cellulose existed in the exploded solid could be decomposed into sugars by an enzymatic reaction. On the other hand, the water-soluble products obtained in the steam-explosion contained C5 sugars. The water soluble products were fed into a hydrothermal membrane reactor in which hydrothermal reaction and pervaporation occur simultaneously to produce furfural (hydrothermal reactive pervaporation). In this presentation, some results of the steam-explosion treatment of rice husk followed by the enzymatic reaction to obtain glucose and the hydrothermal reactive pervaporation to obtain furfural are shown. Also, the total design of the proposed process is examined.

ADVANCED BIOMASS GASIFICATION FOR THE PRODUCTION OF BIOPOWER, FUELS, AND CHEMICALS

Mark A. Paisley, *Taylor Biomass Energy, LLC, 350 Neelytown Road, Montgomery, NY 12549*
Phone: (614) 893-7312, Fax: (614) 459-8579, Email: mark.paisley@taylorbiomassenergy.com

ABSTRACT: Recent price increases for various forms of energy along with projected shortages of supply have resulted in renewed interest in alternative fuels. Biomass gasification provides a renewable basis for supplying not only direct energy products such as gaseous and liquid fuels, and electric power, but also a broad suite of chemicals such as Fischer-Tropsch liquids as well as hydrogen. A medium calorific value (MCV) gas is necessary to achieve the full potential of biomass gasification for fuels, chemicals, and hydrogen production. The Taylor gasification process, being developed by Taylor Biomass Energy is a biomass gasification process that produces a MCV gas. The Taylor gasification process provides improvements over currently available gasification processes by integrating improvements to reduce issues with ash agglomeration and provide in-situ destruction of condensable hydrocarbons (tars), an essential element in gas cleanup. The gas conditioning step integrated into the Taylor Gasification Process provides a unique method to convert the tars into additional synthesis gas and to adjust the composition of the synthesis gas to significantly increase its hydrogen to carbon monoxide ratio. Testing has shown that approximately 90% of the tars can be removed by the gas conditioning step providing a synthesis gas suitable for a variety of applications.

Keywords: gasification, integrated gasification combined cycle, biomass conversion

INTRODUCTION

Biomass resources currently supply over 3 quads to the nation's energy supply and are projected to provide between 17 and 55 quads of the nation's energy needs in the future. A primary benefit of this growth is the potential to increase energy supplies with a renewable, environmentally attractive feedstock. This benefit coupled with the potential to offset recent price increases for various forms of energy and help to impact projected shortages of supply have resulted in renewed interest in alternative fuels from biomass.

Much of the emphasis on biomass supplies has been focused on the production of dedicated biomass crops such as switch grass, woody feedstocks, and other agricultural based materials. To actively pursue commercial scale biomass based systems, however, an adequate, sustainable supply of appropriate biomass material is necessary. Biomass as a fuel source is widely obtainable throughout the United States, but, in many locations, is prohibitively expensive due to transportation costs, seasonality, or local supply issues.

Biomass is an underutilized, renewable resource that can potentially reduce fossil fuel dependency as a part of the primary energy supply. Historically, biomass has merely been used for heating. Today's developing energy technologies are broadening the uses of biomass well beyond simple combustion for heating. The flexibility of biomass as an input to conversion processes provides the ability to generate a wide range of energy products. These include gaseous products that can substitute directly for natural gas or liquid products such as ethanol or Fischer-Tropsh liquids. In addition other products ranging from fuel gas to steam, electric power, and, through synthesis applications, a suite of chemical intermediates and products, or hydrogen can readily be produced from gasification-generated synthesis gas.

To actively pursue commercial scale biomass based systems, however, an adequate, sustainable supply of appropriate biomass material is necessary. Biomass as a fuel source is widely obtainable throughout the United States, but, in many locations, is prohibitively expensive due to transportation costs, seasonality, or local supply issues.

However, if an appropriate biomass feedstock could be recovered from the abundant, primarily biomass, residues from construction and demolition (C&D) and municipal solid waste (MSW) the resources for available, cost competitive biomass could be greatly expanded and, therefore, the supply of bioproducts derived from this resource. Reasonable supplies of such post-sorted MSW and C&D material with all the non-biomass components removed can be derived from the residual materials discussed above, but conventional approaches to energy recovery from these wastes result in environmental problems due to the inherent properties of the technologies themselves.

A FLEXIBLE CONVERSION TECHNOLOGY

Biomass gasification provides a renewable basis for supplying not only direct energy products such as gaseous and liquid fuels, and electric power, but also a broad suite of chemicals such as Fisher-Tropsch liquids as well as hydrogen. Biomass provides the only non-fossil fuel route for renewable hydrogen production. A number of biomass gasification processes have been developed by the US Department of Energy and as well as foreign industry and governments. However, most of these gasification processes are not well suited for the wide range of applications needed to fully exploit the potential of biomass gasification due to their production of a nitrogen diluted, low calorific value product gas.

A medium calorific value (MCV) gas is necessary to achieve the full potential of biomass gasification for fuels, chemicals, and power production. MCV gas can be used directly as a substitute for natural gas in combustion devices such as boilers or turbines but, unlike LCV gas, has the added advantage of being useful for synthesis applications for the production of liquid fuels or hydrogen as shown in Figure 1.

The Taylor gasification process being developed by Taylor Biomass Energy is a biomass gasification process that produces a MCV gas. MCV gas, unlike the low calorific value gas produced in air blown gasifiers, provides the ability to achieve the goal of a true biomass refinery capable of producing not only a fuel gas or power but a range of other products by using the gas as input to a synthesis reactor.

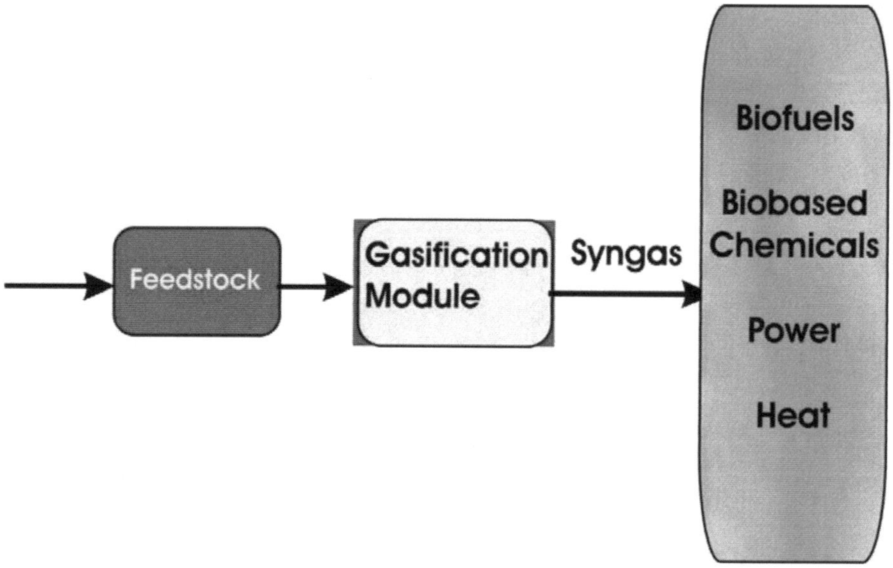

Figure 1. Medium Calorific Value Gas Provides Bioenergy Product Flexibility

BACKGROUND

Of significant concern are supplies of the premium fuels such as natural gas and distillate oil as these fuels both provide environmental advantages to end users as well as convenience for residential use as heating fuels. Importation of both petroleum and natural gas from foreign countries is rapidly increasing. Such imports contribute to the increasing negative balance of trade in the United States. Furthermore, such dependence increases the risk of a foreign nation holding the U.S. hostage as energy security declines.

Biomass provides a means to address these issues when converted in a gasification process, such as the Taylor gasification process, by providing a range of flexible energy products. When a non-nitrogen diluted gas is produced, the options for end use of the synthesis gas greatly expand as illustrated in figure 1 above. The primary emphasis during development of biomass gasification technologies, however, has been toward fuel gas or power. Such emphasis, however, ignores a much broader applicability of the technology.

There are a number of technical issues that must be overcome in order to realize this potential. Many of these issues must also be addressed in order to implement gasification as a part of advanced power systems such as integrated gasification combined cycle (IGCC) cycles. These include, primarily, gas cleanup and conditioning and biomass fuel flexibility. For end use as a synthesis gas, the additional issue of hydrogen content (or the ratio of hydrogen to carbon monoxide in the synthesis gas) must be addressed.

Medium calorific value gas is uniquely suited for use as a synthesis gas for the production of chemicals or hydrogen. By eliminating nitrogen from the synthesis gas, downstream reactor systems to produce these products can approximate their counterparts in petroleum refineries. Even with a medium calorific value gas, the contaminants contained in the synthesis gas produced, consisting of, primarily, condensable hydrocarbons (tars) must be removed. These materials restrict heat recovery from the gases and can cause fouling of downstream equipment or poisoning of synthesis catalysts. One method for avoiding this problem has been to use the gas hot in a boiler or other similar combustion device. Such end use, however, restricts the potential efficiency of such systems and virtually eliminates both the use of high efficiency power production via gas turbines and the use of the gas for synthesis. Other established methods, used for many years in the gas making industry, use wet scrubbing systems that can remove solid particles, soluble gases, soluble salts, and condensable liquids, to produce high levels of fuel purity. These systems, however, in order to achieve high removal efficiencies, utilize solvents to dissolve the aromatic organics present. This then leads to disadvantages related to disposal of contaminated water. These systems also introduce additional heat losses which can adversely affect overall system efficiencies.

Taylor Biomass Energy has developed an advanced, indirectly heated gasification process that effectively converts the tars in the gas to non-condensible, lower molecular weight species. This removal allows a higher level of the sensible energy contained in the synthesis gas to be recovered while simplifying any secondary conditioning of the gas that might be necessary. By building on the success of other indirectly heated gasification processes, the Taylor Gasification Process provides a means to utilize the broad range of biomass feedstocks that are available while simplifying the process of converting these biomass feedstocks into economically viable energy products.

THE TAYLOR GASIFICATION TECHNOLOGY

The Taylor Gasification Process, being developed by Taylor Biomass Energy is a biomass gasification process that produces a medium calorific value gas having a higher heating value approximately half that of natural gas. This gas, unlike the low heating value gas produced in air blown gasifiers, provides the ability to directly substitute for natural gas or be used as a synthesis gas for the production of ethanol or, alternatively, for the production of power, chemicals, or hydrogen.

Indirectly heated gasification has been demonstrated to be a flexible and reliable method for efficiently producing a medium heating value gas from biomass based feedstocks. A number of technologies have been successfully demonstrated at commercial scale that utilize this type of conversion process. These include the FICFB gasifier in Gussing, Austria; the SilvaGas gasifier in Burlington, VT; the ENSYN pyrolysis process; and the Thermochem process along with other processes at earlier stages of development at various institutions in Europe and China. The Taylor gasification process, with a patent pending, builds upon the results of these technologies. Taylor's process, however, provides improvements in operation by integrating innovative improvements to the technologies to reduce issues with ash agglomeration and in-situ destruction of condensable hydrocarbons, an essential element in gas cleanup and environmental performance of the process. By providing in-situ removal of the condensable hydrocarbons, subsequent gas cleanup is greatly simplified as the gas may be cooled prior to final cleanup thus reducing the size of the cleanup equipment. Such cooling improves overall process efficiency by providing a means to more effectively recover sensible energy in the product gas stream.

In the Taylor Gasifier, a circulating heat carrying material is used to rapidly heat the incoming biomass and convey unconverted materials from the gasification reactor into an associated combustion reactor, as shown in figure 2 below.

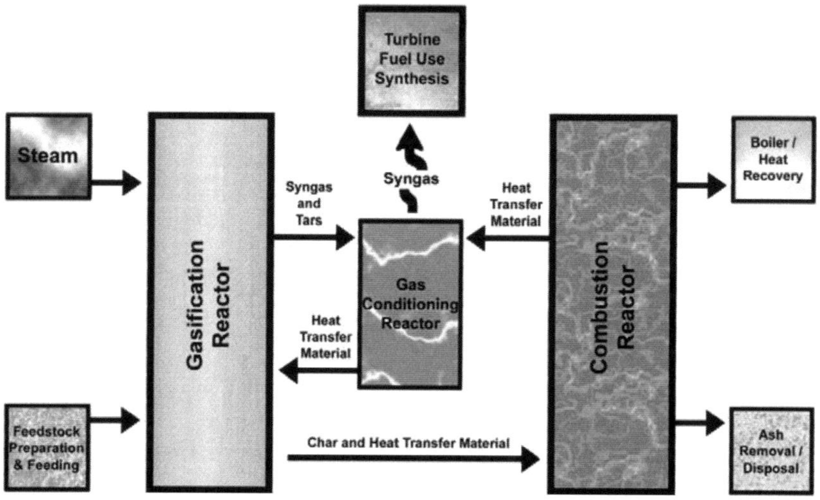

Figure 2. Block Diagram of the Taylor Gasification Process

In the gasifier, biomass is contacted by the heated heat carrying material, and steam. No air or oxygen is added so there are no combustion reactions taking place, providing environmental advantages. The biomass is rapidly (less than one second) converted into medium calorific value gas (14-17 MJ/Nm^3) at a temperature of approximately 850C. Any unconverted material along with the cooled heat transfer material, pass through the gasifier and then are separated from the product gas. The product gas continues on to the gas conditioning step prior to any final gas cleanup that might be needed while the solids are conveyed into the process combustion reactor.

In the combustion reactor, air is introduced which consumes the char and, in the process, reheats the sand to approximately 1000C. In the combustion reactor all remaining carbon is consumed, resulting in a carbon-free ash. Due to the combustion conditions and the fact that the unconverted material is essentially carbon, emissions are low from this step in the process. The reheated solids are separated from the flue gas and returned to the gasification reactor. Ash is removed from the flue gas, resulting in a high temperature (1000C), clean gas stream available for heat recovery.

The gas conditioning reactor is the key element of the Taylor Process that provides enhanced gas compositions along with the improved heat recovery potential. Within the gas conditioning reactor, the product gas contacts the high temperature solids (1000C) providing an optimum environment for steam reforming of the tars. The tars are converted to lower molecular weight compounds that augment the quantity of synthesis gas produced.

The additional residence time provided by the gas conditioning reactor in the presence of a catalytic medium (the hot circulating solids) allows the synthesis gas to reach water gas shift equilibrium. As a result the hydrogen content of the synthesis gas is enhanced compared to other biomass gasification processes as shown in Table 1 below. The steam reforming reactions and the water gas shift reaction are balanced thermally so no cooling of the circulating solids takes place.

Table 1. Comparison of Taylor Syngas Compositions With Other Indirect Gasification Processes

Component, o/v	Taylor	FICFB	SilvaGas®
Hydrogen	45-48	37.7	20.7
Carbon Monoxide	15-20	29.1	46
Methane	10-13	10.4	15.6
Ethylene	1-3	2.8	5.3
Carbon Dioxide	18-20	19.6	11.1
Ethane	0-1	0.3	0.7
Nitrogen	trace	0.1	0.6

As an additional benefit, the syngas exits from the process at a higher temperature (1000C) and, because of the significant reduction in condensable materials, can be cooled to a much lower temperature resulting in an increase in heat recovery of nearly 40% when compared to syngas generated by other indirectly heated gasification processes.

DEVELOPMENT OF THE PROCESS

An extensive data base exists on indirectly heated gasification processes. This data was generated by Repotec in Gussing, Austria, by Battelle in Columbus, Ohio, by the National Renewable Energy Laboratory in Golden, Colorado, and by other research laboratories throughout the world. Because of the information and data on both the gasification and combustion reactors in the process contained in this data base, only the performance of the unique gas conditioning reactor had to be investigated by Taylor Biomass. To confirm the performance of this gas conditioning step, a series of tests were conducted at the National Renewable Energy Laboratory in their Thermochemical Users Facility (TCUF).

This unit, while not in the exact configuration of the Taylor gasification process can be used to accurately simulate the process to allow determinations such as those for this study to be conducted. A description of the facility follows:

Thermochemical Process Development Unit

The Thermochemical Process Development Unit (TCPDU) is an integrated system of unit operations designed to investigate biomass thermochemical conversion to gaseous and liquid fuels and chemicals. The individual unit operations were designed to permit multiple equipment configurations. The ability to reconfigure the TCPDU permits operation over a wide range of conditions.

In the process configuration for gasification, the first and primary reactor in the process is an electrically heated 20.3 cm diameter fluidized bed reactor with a 40.6 cm (16") diameter freeboard. Product gas, entrained char, and bed material flow from the reactor through a pipe into a thermal cracker. The thermal cracker ensures that the raw composition of the product gas along with tars and other contaminants is appropriate for the gasification process being examined. Downstream of the thermal cracker are two cyclone separators in series that remove solids from the gas. The gas leaving the cyclones moves quickly into a "full stream tar reformer."

The Full Stream Reformer (FSR) shown in figure 3 below, is an electrically heated, 35.6 cm (14") diameter fluidized bed reactor with a 61 cm (24") diameter freeboard. Process gas from the gasification system is mixed with superheated steam or nitrogen as required and is then heated up to or near bed temperatures in a preheater. The reformed gas exits the FSR via an internal 11.4 cm (4.5") diameter cyclone. Entrained solids from the catalyst bed are collected by the cyclone and returned to the bed through a 4 cm (1.5") diameter dip leg. For this series of experiments, the full stream tar reformer was used to evaluate the Taylor "gas conditioning reactor".

This experimental system is outfitted with extensive analytical measurement devices. These include on-line gas chromatography and a transportable molecular beam mass spectrometer (TMBMS) to provide a continuous measurement of the process performance including concentrations of tar species present before and after the tar cracking operation.

Overview of the TMBMS

The TMBMS was used to provide continuous, real-time chemical analysis of the hot syngas at the inlet and outlet of the full-stream reformer. The hot process gas arrives at the entrance of the instrument by means of a heated sampling system, which will be described below. A molecular beam forms as the gases are extracted through a heated 300-μm critical-flow orifice into a three-stage, differentially pumped vacuum system. The gases then undergo a free-jet expansion, essentially halting chemical reactions and preventing condensation. Components of the molecular beam are then ionized by 22.5 eV electrons before passing through a quadrupole mass analyzer. The ions are then detected with an off axis electron multiplier and mass spectra are generated according to the molecular weight of the ions. TMBMS data collection and the mass spectrometer were automated using a PC-based data acquisition and control system. Mass spectra were recorded for m/z=3 to m/z=400 along with auxiliary pressure data at the rate of 36 scans/min with 1 minute averages stored. A schematic of the TMBMS system is shown in figure 4.

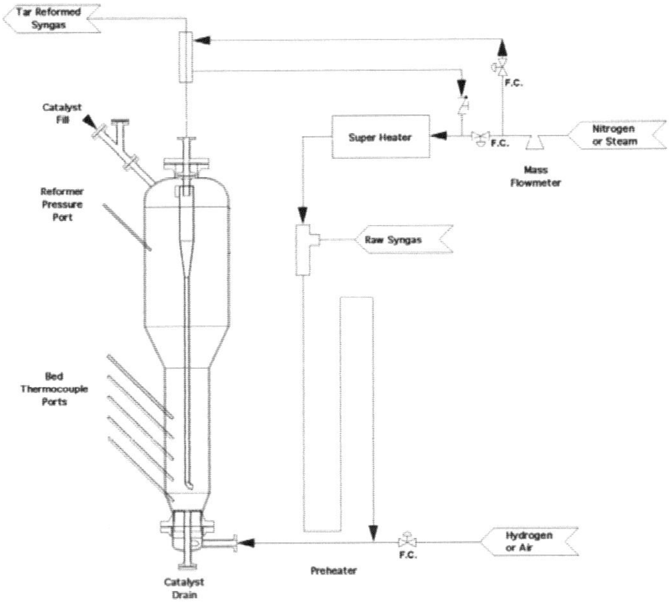

Figure 3. Full Stream Tar Reformer

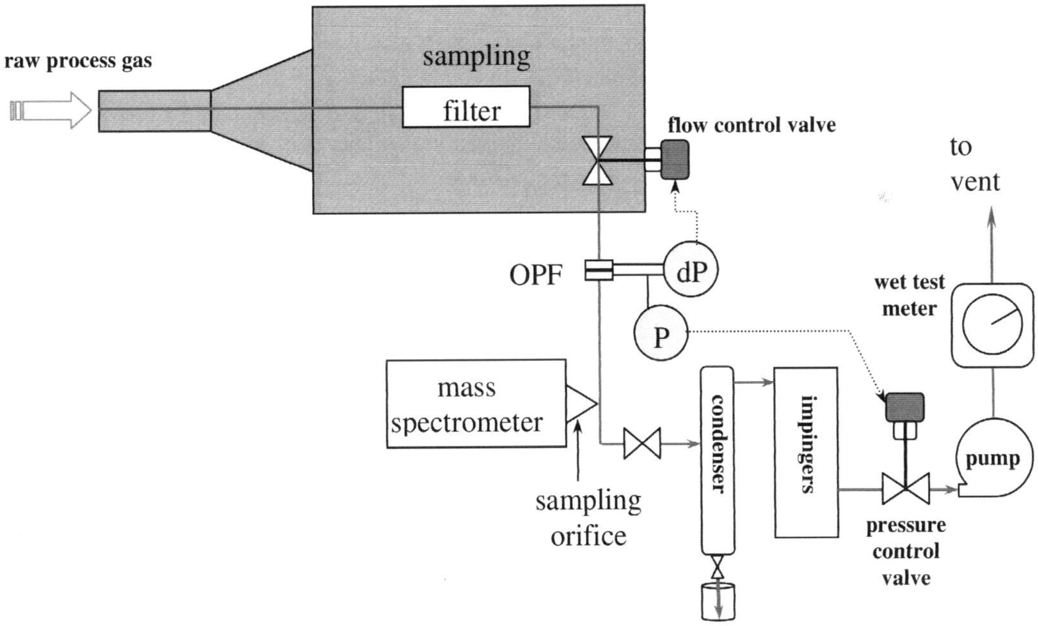

Figure 4. Schematic of TMBMS System

For these tests, NREL's TCPDU was operated in indirect gasification mode using steam and nitrogen as the fluidizing gas/oxidizer to produce a medium calorific value synthesis gas from the incoming biomass feedstock. In order to provide process stability, all critical process parameters for the TCPDU, including temperatures, pressures and flow rates, are automated by a PC-based data acquisition and control system. These parameters were generally stable to within a few percent over a given experimental run. In the full stream reforming reactor, the temperature in the fluidized bed was maintained at 925°C.

EXPERIMENTAL RESULTS

Based on data measured using the TMBMS and on-line gas chromatography, tar conversion results showed between 80 and 90% conversion of nearly all species present. An increase in benzene was noticed, indicating that a portion of the tars was converted to this stable compound. This is not an unexpected result as the tars are primarily aromatic in nature and benzene is a very stable aromatic molecule. Figures 5 and 6 illustrate the tar compounds found in the inlet and outlet of the full stream tar reformer.

Water Gas Shift Results

As discussed above, an additional benefit of the gas conditioning reactor, is the resulting increase in hydrogen content due to water gas shift equilibrium and the conversion of the tars into additional syngas. Table 2 shows the inlet and outlet composition of the synthesis gas illustrating this desirable result.

Such a change in composition allows the syngas from the Taylor gasifier to be used directly for a much broader range of applications including the use for synthesis applications or for the direct recovery of hydrogen as an end product.

Table 2. Comparison of Gas Compositions at the Inlet and Outlet of the Gas Conditioning Reactor

Component, o/v	Inlet	Outlet
Hydrogen	29.5	47.2
Carbon Monoxide	30.1	17.4
Methane	14.4	12.2
Ethylene	3.9	2.6
Carbon Dioxide	20.7	20.1
Ethane	0	0.1
Nitrogen	trace	trace

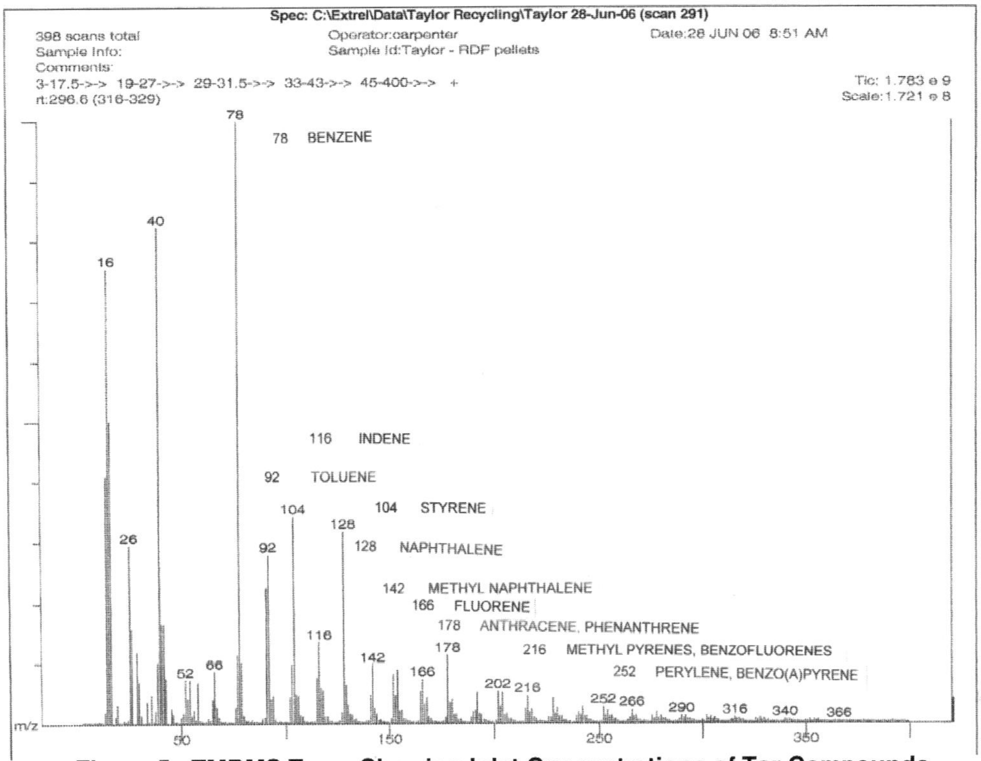

Figure 5. TMBMS Trace Showing Inlet Concentrations of Tar Compounds

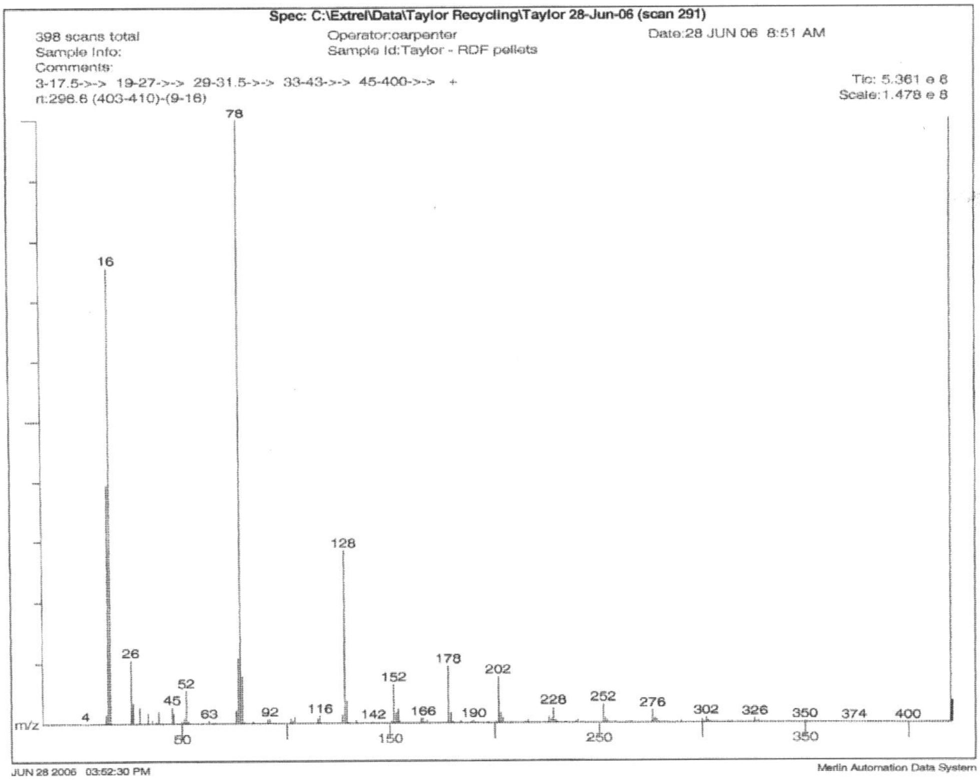

Figure 6. TMBMS Trace Sowing Outlet Concentrations of Tar Compounds

Lifetime of Catalytic Performance

When applications rely upon the catalytic performance of a material, as in the Taylor gas conditioning reactor, there is always a question about the expected lifetime of the catalyst material. One primary cause of this deactivation in steam reforming type of reactions is the formation of carbon deposits on the catalyst surface. As the catalytic material in the Taylor Gasification Process is the circulating heat transfer material, such deactivation can not occur as the material passes alternately through an oxidizing regime and a reducing regime. This allows for in-situ regeneration of the catalyst thereby virtually eliminating the problem of catalyst deactivation.

During the testing at NREL, this "regeneration" of the material was evaluated. The "regenerated" material showed essentially the same tar conversion activity as the "virgin" material. This is illustrated in Figure 7 which shows the conversion performance on two consecutive days of operation. A "regeneration" of the material had taken place between the two testing periods.

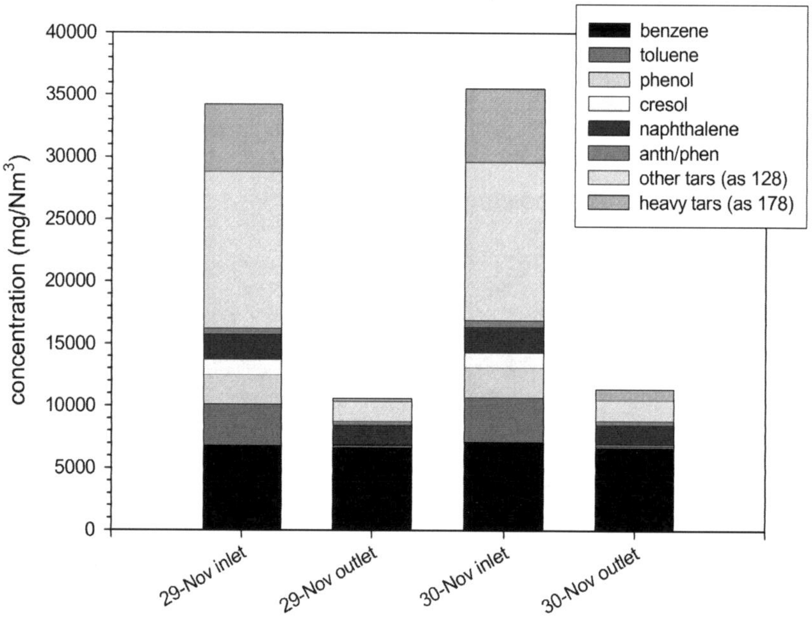

Figure 7. Consistent Performance of the Gas Conditioning Step

COMMERCIAL APPLICATION OF THE TAYLOR GASIFICATION PROCESS

To actively pursue commercial scale biomass based systems, an adequate, sustainable supply of appropriate biomass material is necessary. Biomass as a fuel source is widely obtainable, but, in many locations, is prohibitively expensive due to transportation costs, seasonality, or local supply issues.

The disposal of MSW is becoming an increasingly serious problem throughout the United States. In the period 1990 to 2000, MSW generation rates have increased by 13 percent. When compared to 1980 levels, generation rates have increased over 50 percent. Large quantities of other solid wastes, commonly referred to as construction and demolition wastes (C&D) are also generated annually further increasing the problem.

Taylor Biomass Energy is constructing a facility at its site in Montgomery, NY to utilize biomass recovered from these plentiful and sustainable materials and convert them into green bioproducts. This integration of the Taylor sorting, separating, and recycling process along with the Taylor Gasification Process is underway.

At the site currently, approximately 275 tonnes per day of C&D material along with 90 tonnes of wood residues (stumps and similar materials) are sorted and a significant portion of biomass that would be suitable input to the gasification process is recovered. This existing operation will be expanded by approximately 50% and a sorting operation to handle approximately 450 tonnes per day of municipal solid waste is being added. The combined operation will produce, on a sustainable basis, approximately 275 tonnes per day (dry basis) of a "processed biomass fuel" for the gasification process.

A schematic showing the process arrangement for the Taylor gasifier is found in figure 8. In the diagram, the flow of the hot solids through the gas conditioning reactor on their return to the gasification reactor can be clearly seen. The cleaned gas is then directed to heat recovery to enhance the overall process efficiency.

The Taylor gasifier will be constructed in a modular fashion rather than being completely assembled on site. This fabrication technique will provide much tighter control of the assembly as a well as provide capital cost savings during construction. Completed modules will be delivered to the site and erected with a minimum of installation necessary on site. A 3-D rendering of the assembled gasification process is found in Figure 9.

Syngas generated from the incoming biomass will be cleaned of particulates, compressed and used as fuel for a gas turbine based combined cycle power system. Approximately 20 MW will be exported from the facility to a local utility substation located on the site.

In addition, the facility will be constructed to provide a development platform where additional feedstocks and / or downstream unit operations such as chemical synthesis or hydrogen production can be readily demonstrated.

The Montgomery facility is underway with the Class A design package completed, permit documents in review, project financing in place, and initial site preparations completed. Detailed design is underway and construction of the gasification modules is expected to start shortly.

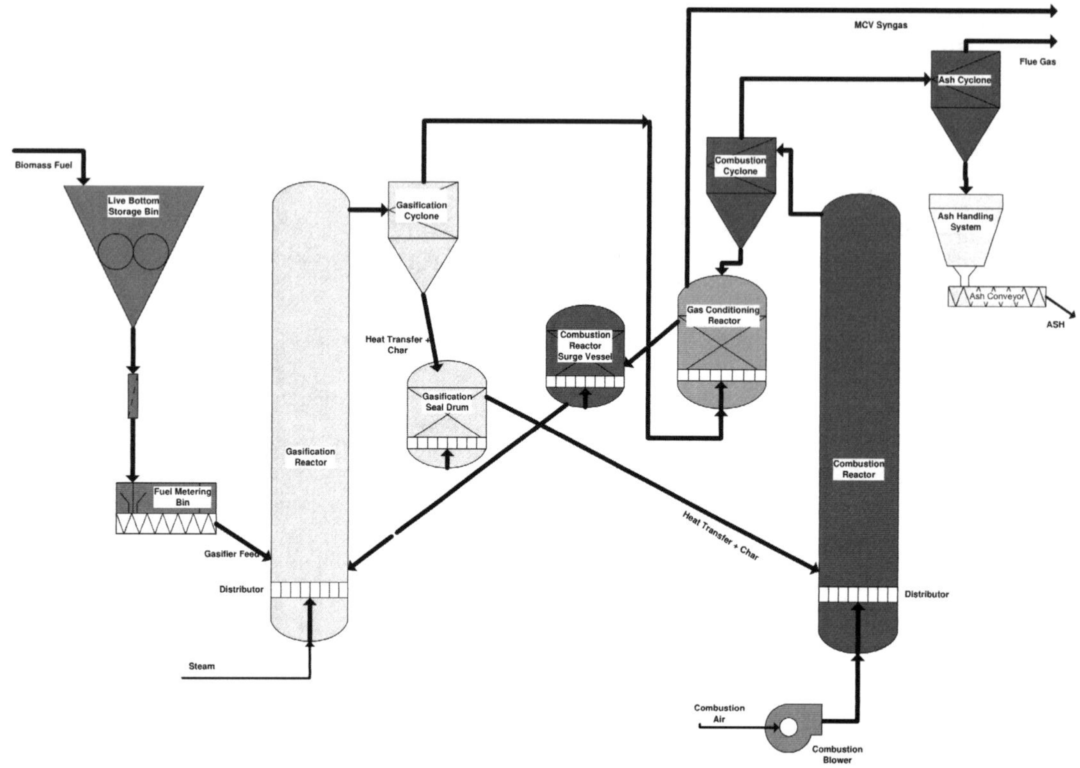

Figure 8. Schematic Diagram of the Taylor Gasification Process Showing Process Arrangement

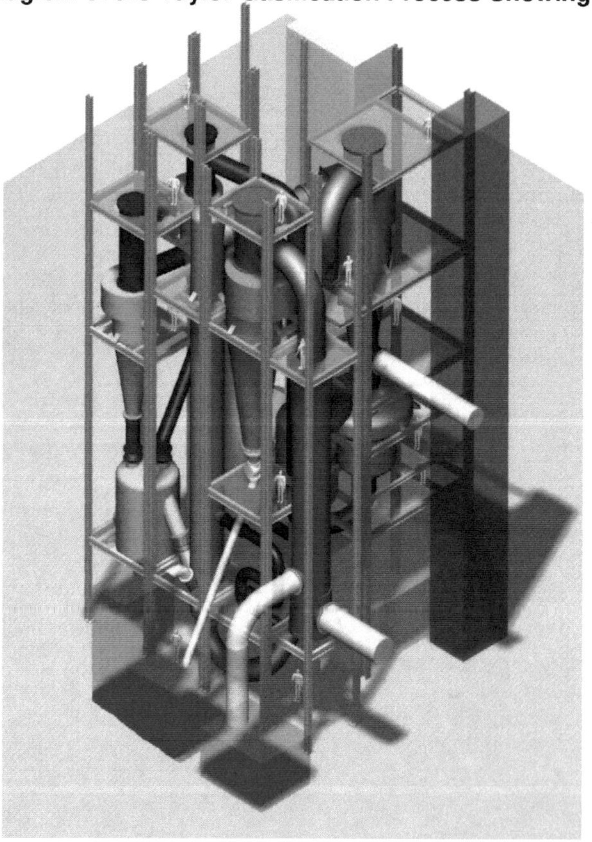

Figure 9. Layout of the Taylor Gasification Process Modules

PROCESS ECONOMICS

The installed capital cost for the gasification island has been estimated to be between \$550 and \$600 per installed kW. Final costs for the power island have not been determined, but are expected to be consistent with state-of-the-art combined cycle facilities based on natural gas firing.

The facility at the Montgomery site will provide approximately 20 MW of power to the New York grid. This power is anticipated to be considered "renewable power" and therefore will be able to take advantage of various incentives available at the State and National level.

Additional income to the project will accrue based on the sale of recyclable materials recovered from the incoming C&D and MSW streams and from tipping fees associated with the handling of these materials at the site.

Overall, the project is expected to be profitable in the very early years of the project life.

CONCLUSIONS

The Taylor gasification process has been shown to provide enhanced conversion of condensable organics by the integration of a unique gas conditioning step within the process. This step has been experimentally verified through testing at NREL and shows significant advantages when compared to downstream unit operations.

The commercial demonstration of the integrated Taylor sorting, separating, and recycling process with the Taylor gasifier has also shown that:

- A viable biomass resource can be supplied from residues including C&D materials and MSW
- Incorporation of these residue based biomass resources can be a key element in economical biomass energy projects
- A high quality synthesis gas can be produced from biomass without the need for multiple downstream process steps
- Biomass gasification can provide a route to a wide range of energy products including power, liquid fuels, chemicals, and hydrogen

Additional projects employing the Taylor gasifier and the Taylor sorting, separating, and recycling processes are in various stages of development. These projects are expected to be "on-line" within one year after the completion of the Montgomery, NY project.

REFERENCES

1. Marquard-Mollenstech, T., et al, "New Approach for Biomass Gasification to Hydrogen," 2nd World Conference and Technology Exhibition on Biomass for Energy, Industry and Climate Protection, Rome, Italy, 10-14th May 2004
2. Reed, T., and Gaur, S., "A Survey of Biomass Gasification 2000," BEF Press, 2001
3. Paisley, M.A., and Overend, R., "Verification of the Performance of Future Energy Resources' *SilvaGas®* Gasifier Operating Experience in The Vermont Gasifier", Pittsburgh Coal Conference, 2002
4. Paisley, M.A., "Biomass Energy," Kirk-Othmer Encyclopedia of Chemical Technology," John-Wiley and Sons, 2003

Influence of Moisture Content of Woody Biomass on Heat Transfer with the Heterogeneous Chemical Reactions Caused in the Pyrolysis

Ken-ichiro Tanoue, Thaung Oo, Tatsuya Hinauchi, Tatsuo Nishimura,
Yamaguchi University, Ube, Japan

Miki Taniguchi and Ken-ichi Sasauchi,
Chugai Ro Co.ltd, Sakai, Japan

Introduction

Pyrolysis of biomass consists of many interdependent reactions; nevertheless, it can be reduced to the reaction, which is universally known as the Broido-Shafizadeh mechanism [2,11]. In the primary reactions, biomass decomposes to gas, tar and char. This reaction has been studied extensively in the form of cellulose decomposition. In the secondary reaction, the tar generated in the primary reaction decomposes to light gas and char. However, as the evaporation process is not included in the mechanism, it is difficult to understand whether the decomposition is the endothermic process or the exothermic one.

Pyrolysis rates in woody biomass have been investigated using various methods, including thermogravimetric analysis [5,7,11,13], analysis of weight change among the main components of the product mixture (gas, tar and char) [3,4,10], and analysis of heat and mass transfer during the reaction [1,6]. In all of these studies, the kinetics of pyrolysis were analyzed using samples with weights of particle diameters of a few tens of microns in diameter. However, in a real pilot plant, sample sizes in grams and biomass particles with an average diameter of several mm have been utilized 8). In order to estimate the pyrolysis rate based on the results obtained [1, 3-7, 10, 11, 13], an understanding of heat transfer not only in the biomass particles but also in the biomass layer is required.

In our previous report[9], transport phenomena in the rapid pyrolysis of woody biomass was investigated experimentally. From the results, the pyrolysis depends strong on the setting temperature and can be divided into three processes; i.e., i) evaporation of water, ii) decomposition of biomass and iii) decomposition of generated char.

The aim of this study is to investigate the effect of moisture content of biomass on the slow pyrolysis process. The effect of moisture content in the biomass on the slow pyrolysis process has been investigated by using fixed gasifier. To be more concrete, we especially focused on average temperature and heat flow rate in biomass layer for heat transfer while we focused on generated gas flow rates and concentration of generated gases for mass transfer.

Experimental procedure

Figure 1 shows the apparatus used for pyrolysis experiments with woody biomass. The rig consisted of a nitrogen gas supply, a tubular reactor, cold traps for tar and water, a gas-sampling bag, and a gas flow meter. The furnace had a maximum output of 1.5 kW, and the wall temperature of the reactor could be set to temperatures up to about 800ºC. The cold traps, which were situated in an ice bath, were two 500-mL Erlenmeyer flasks filled with glass wool and solid $CaCl_2$. During the reaction, tar and water were adsorbed by the traps, and the gas generated in the reaction flowed through the traps and filled the sampling bag (GL Science Co., Ltd.) for measurement of the gas component. Gas flow rate was measured using a wet-type gas meter (Shinagawa Co., Ltd., W-NK-2B). An O_2 monitor was placed at the exit of the reactor. Before the experiment, air in the reactor was replaced by nitrogen gas, which was allowed to flow into the reactor until the concentration of O_2 at the exit was less than 1%, after which time the nitrogen gas supply was stopped.

Woody biomass particles (sample weight W_0 = 100 g, Dp,50 = 1.1 or 11 mm) were introduced into the tube reactor, and the wall of the reactor was heated to 800°C by the electric furnace at heating rate of 400°C h^{-1}. The temperature of the reactor wall was then maintained at 800°C for about 60 min. Two samples with different particle sizes were used, and for each sample, two pyrolysis experiments were conducted: in the first, the temperature profile of the biomass layer, the flow of gas generated, and the material balance were measured; and in the second, the components of the generated gas were analyzed.

There are nine points at which the temperature was measured in the biomass layer. The temperature both in the biomass layer and in the reactor wall was measured at one-minute intervals, and the heat flow rate was calculated based on the change in the average temperature of the biomass layer over 1 min.

The carbide material remaining in the reactor is defined as char. Samples of the gas generated in the reaction were collected in the sampling bag every 20 min and analyzed quantitatively by gas chromatography (GL Science Co., Ltd., GC323) to determine the concentrations of H_2, CH_4, CO and CO_2. The experimental conditions are shown in Table 1.

Results and discussion

Heat transfer in the biomass layer

Figure 2 shows the propagation of temperature change in the biomass layer. Initial temperatures in the layer are almost uniform at 32°C ± 3°C. Between t = 15 min and t = 45 min, heat transfer is controlled by conduction from the furnace to the biomass layer. At t > 60 min, radiation from the furnace to the upper surface of the biomass layer begins to influence heat transfer. At 75 min < t < 90 min, the

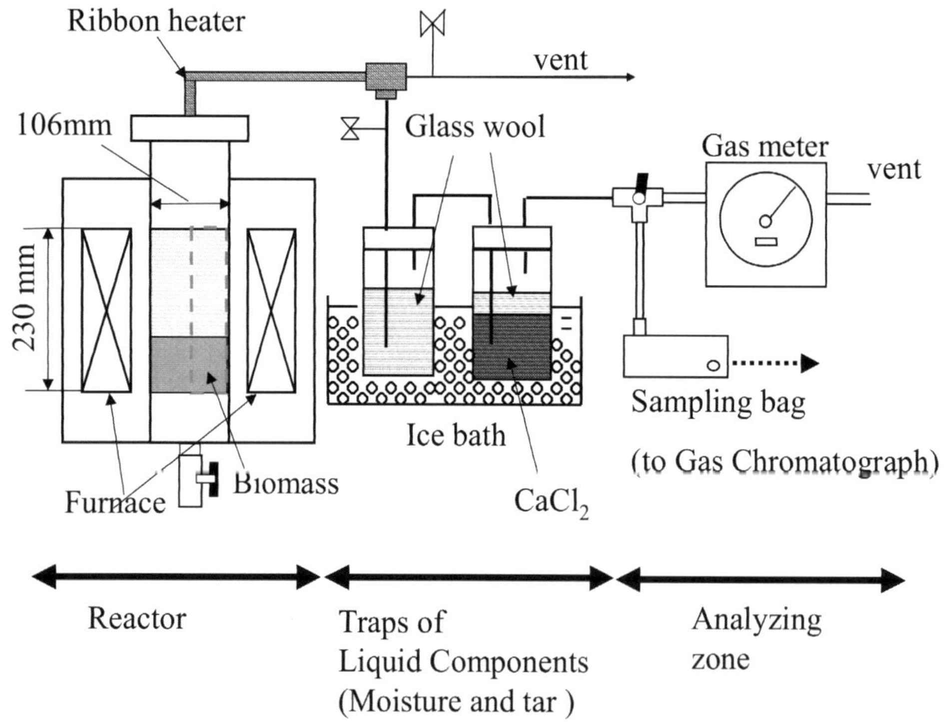

Figure 1. Experimental test rig for the pyrolysis of woody biomass.

temperatures in the layer increase quickly, and almost all of them are greater than 400°C. Finally, at t > 120 min, temperatures in the biomass layer exceed 700°C.

Figure 3 shows the average temperature in the biomass layer as a function of time. The solid line shows the set temperature of the electric furnace at r = 53 mm and z = 110 mm. At t < 60 min, the average temperature increases slowly compared with the increase in the set temperature. At 60 min < t < 75 min, the temperature begins to increase more quickly, and at 75 min < t < 120 min, the increase is linear with

Table 1. Experimental conditions.

	Reactor	
Heating rate	$\dot{T}_S$ (°C/hr)	400
Temperature set for the furnace	T_S (°C)	800
Operation time	t (min)	180
	Biomass	
Average particle size	$D_{p,50}$ (mm)	1.1, 11
Initial weight	W_0 (g)	100
Weight % of additional water	φ (%)	0, 30, 60

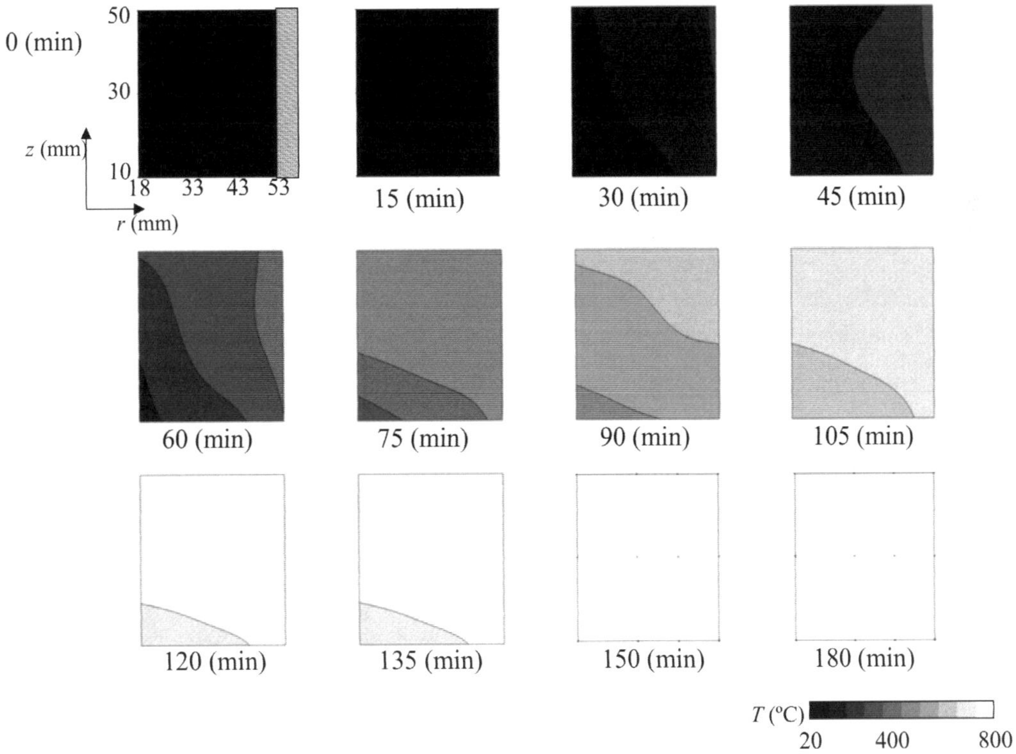

Figure 2. Change of the temperature distribution in the biomass layer as a function of time through the pyrolysis.

time. Finally, at t > 120 min, the average temperature approaches a constant value. The results obtained using samples with different average particle sizes showed that the average temperature did not change even when the particle size was varied by ten times.

Heat flow rate of inside biomass layer for different moisture as a function of time are shown in Figure 4. Figure 4a shows for the heat flow rate of wet 0% biomass. Between 0-60 min, heat flow rate increases to heat source line (6.6℃/min) as average temperature of biomass layer is increased to 220℃ and at which evaporation process is occurred inside biomass layer and virgin biomass takes heat from surrounding to evaporation take place. After which, heat flow rate line is rapidly increased to 1st peak at 70 min as the reaction is started to occur and average temperature is rapidly increased 220℃ to 350℃ at which decomposition of cellulose and tar formation are occurred. Exothermic reaction is predominately occurred due to the decomposition of cellulose at that time. After that, temperature gradient line is decreased due to the decomposition of tar is occurred between 350 ~ 500℃ of average temperature at which endothermic reaction is predominately occurred. Above 500℃, temperature gradient line is increased to 2nd peak at 100 min, 600 ℃ due to the decomposition of lignin at which exothermic reaction is occurred again. It can be suggested that 1st peak is caused by the decomposition of cellulose with exothermic reaction and 2nd peak is caused by the decomposition of lignin with exothermic reaction is occurred in biomass layer. Figure 4b shows the heat flow rate line of wet 60% biomass. Between 0-90 min, temperature gradient line is slightly increasing and decreasing at the beginning of experiment and then increases to intersection point of heat source line of furnace and heat flow rate line of wet 60% biomass as average temperature line is increased to 220℃.During this period, evaporation process is occurred inside biomass layer and heat of evaporation is mainly occurred. After which temperature gradient line is rapidly increased to its peak position at 115 min due to the decomposition of cellulose and lignin is occurred simultaneously as average temperature is rapidly increased from 220℃ to 550℃.After that point, heat flow rate line is rapidly decreased and rapidly increased again. Which can be seen in figure 4b and shows with oval shape marking since fast reaction was occurred (ie, reaction is still occurred at higher average temperature of 700℃ in inner part of biomass layer near the center of fixed gasifier). It is suggested that the peak position of heat flow rate line for wet 60% is due to the heat of reaction which is caused by decomposition of cellulose and lignin with exothermic reaction is predominantly occurred.

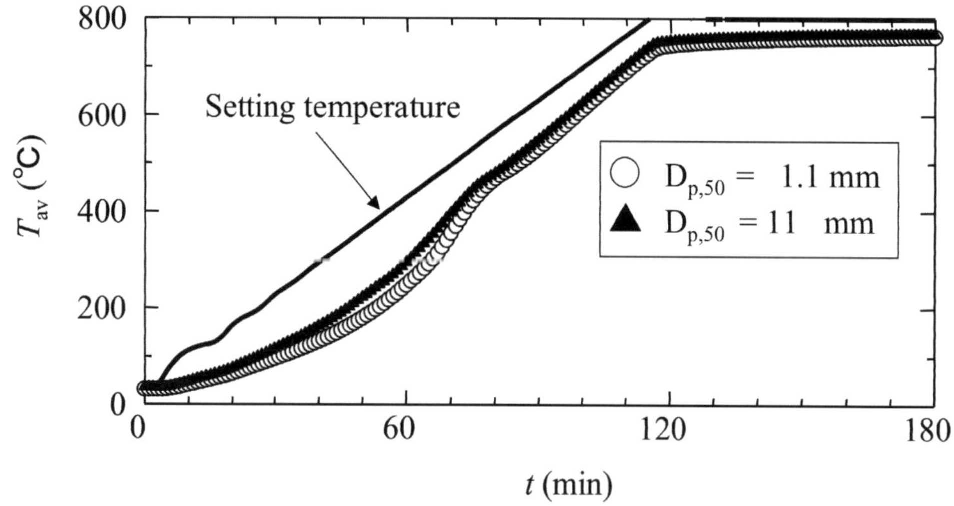

Figure 3. Change of the average temperature as a function of time through the slow pyrolysis.

Generated gas flow rate of product gas

Generated gas flow rates (ml/min.g) of product gas (H_2, CH_4, CO_2 and CO) as a function of time for different moisture content of biomass are illustrated by Figure 5. According to the concentration of generated gases, its flow rates and generated gas in volume percent, it can be estimated the generated gas flow rates of product gases for 1 gram of biomass. The most important reactions during the slow pyrolysis (without combustion) are: $C+H_2O \rightarrow CO+H_2$ (1), $CO+H_2O \rightarrow CO_2+H_2$ (2), $C+2H_2 \rightarrow CH_4$ (3), $CO+3H_2 \rightarrow CH_4+H_2O$ (4), $CH_4+H_2O \rightarrow CO+H_2$ (5). For wet 0% biomass case, any gas doesn' t generate between 0~40 min and only water vapor is developed at evaporation time. Between 40-60 min is transition time, the time between evaporation time and reaction starting time at which a few amount of CO_2 and CO gas are started to generate. Between 60~90 min, high molecular weigh of CO_2, CO, and CH_4 gases are mainly generated as cellulose decomposition occurred and biomass (carbon) is started to convert burnable gases at which reaction 1,2 and 3 are mainly occurred inside wet 0% biomass layer. Between 90-150 min, low molecular weigh of H_2 gas is mainly generated as decomposition of lignin occurred and higher molecular weight of gases are decreased in generating at which reaction 4 and 5 are mainly occurred inside biomass layer. It can be seen that there are two peak positions in generated gases. 1st peak is due to the mainly generating of CO_2, CO and CH_4 gas at 80 min and 2nd peak is due to the mainly generating of H2 gas at 120 min. CO_2 and CO gas are firstly generated and CH_4 gas is generated by little delay in time but H2 gas is lately generated. Generated product gases flow rate of wet 0% figure

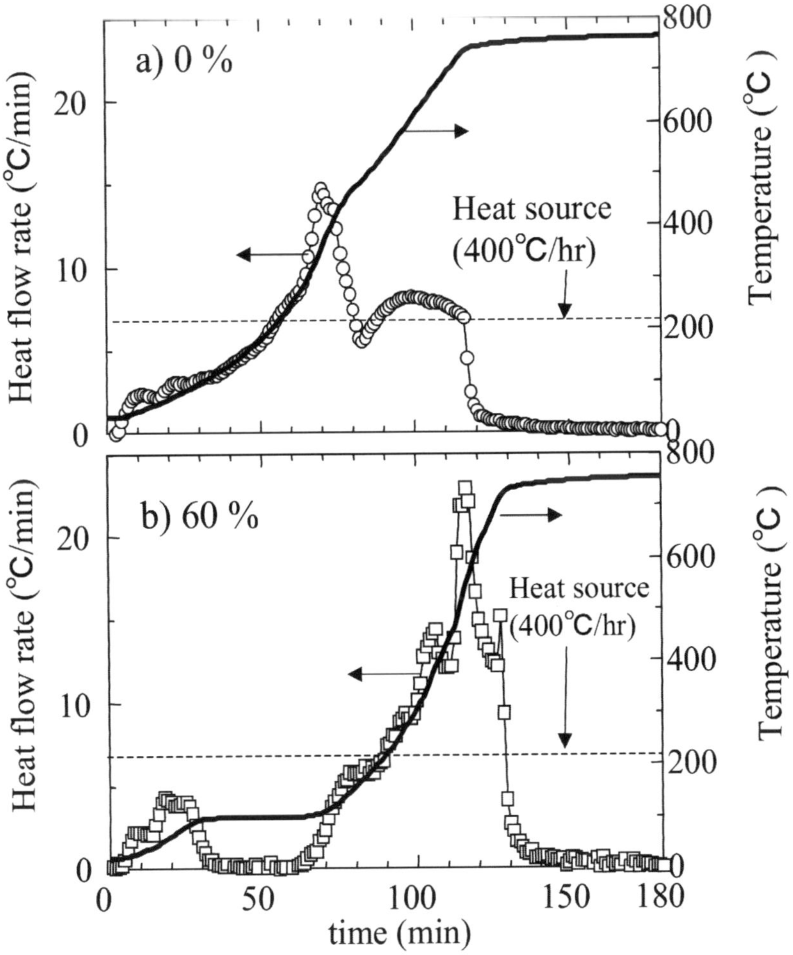

Figure 4. Heat flow rate and temperature inside biomass layer as a function of time.

suggested that higher molecular weight of gas is firstly generated and low molecular weight of gas is lately generated consecutively in slow pyrolysis process. On the other hand, for wet 60% biomass case, any gas doesn't generated between 0~ 60 min and only water vapor is developed at evaporation time. Between 60-90 min is transition time, not only water vapor generate but also some CO_2 and CO gas are started to generated due to the evaporation process still going on and carbon are started to convert. Between 90 ~150 min, H_2, CH_4, CO_2 and CO gases are mainly generated at 120 min as the decomposition of cellulose and lignin is occurred simultaneously at which reaction 1 to 5 are occurred inside biomass layer but reaction 5 is mainly occurred at that time due to the higher in gas flow rate of H_2. It is possibly that still remaining of moisture content (H_2O) inside wet 60% biomass layer and which is future reacted with already generated gas of CH_4 and yields more H_2 gas. High molecular weight of CO_2 is firstly generated and low molecular weigh of H_2 is lately generated as like as wet 0% and wet 30% biomass case.

Conclusions

Relationship between heat transfer and heterogeneous chemical reactions in the biomass layer through the pyrolysis of wet woody biomass was investigated experimentally. The main conclusions may be summarized as follows:

1) The results obtained of pyrolysis showed that the average temperature in the biomass layer increases slowly compared with the rate of change of the set temperature. This tendency did not vary for samples with different average particle sizes (1.1 mm < Dp,50 < 11 mm).

2) For the biomass without adding water (j = 0wt %), by comparing heat flow rate in the biomass layer with the heat source, there are the evaporation of water and two exothermic chemical reactions; the exothermic reactions coincided with two peaks observed in the flow rate of generated gas. CO_2 was

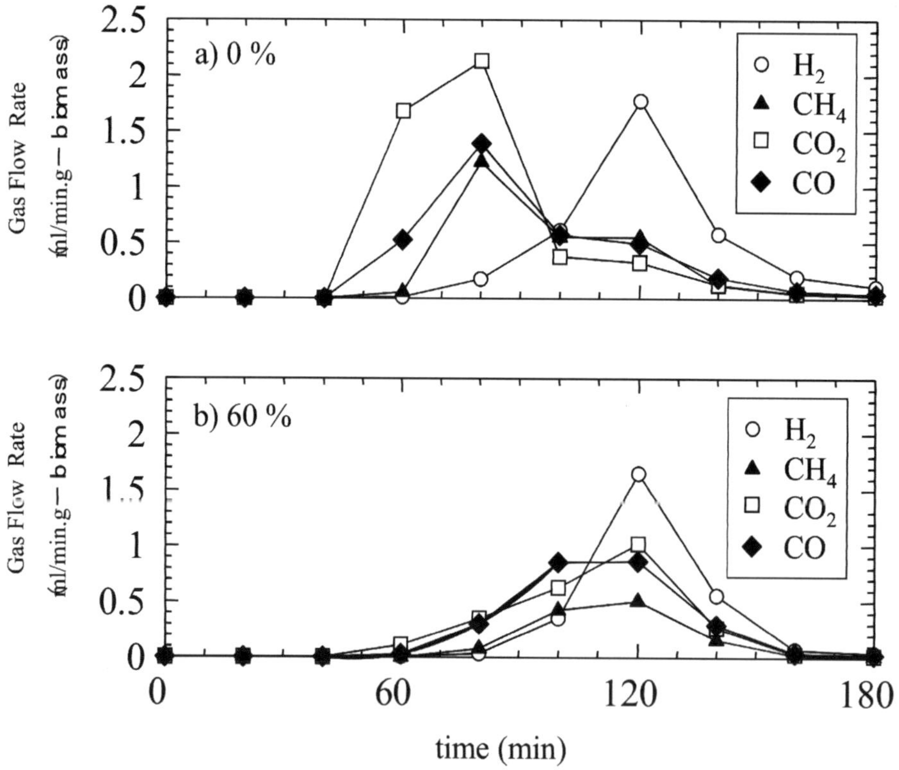

Figure 5. Generated product gases flow rate as a function of time.

3) For the biomass with adding water (j =60 wt %), it was difficult to separate the chemical reactions in j =0wt % because the heat transfer had only one exothermic region except for the evaporation.

Acknowledgements

This work was supported in part by a Grant-in Aid for Scientific Research C (No. 19560174) from the Japan Society for the Promotion of Science. Our special thanks are extended to Dr. Morihisa Yokota of Ube Industries, Ltd., for helpful discussions on the work.

Refernces

1. Babu, B. V. and Chaurasia, A. S. (2004), "Heat transfer and kinetics in the pyrolysis of shrinking biomass particle," Chem. Eng. Sci., 59, pp.1999-2012
2. Broido, A. (1976), "Kinetics of solid-phase cellulose pyrolysis. In: F. Shafizadeh, K. V. Sarkanenm, D. A. Tillman, editors. Thermal uses and properties of carbohydrates and lignins," pp.19-35, New York, Academic Press
3. Fagbemi, L., Khezami, L. and Capart, R.(2001), "Pyrolysis products from different biomasses: application to the thermal cracking of tar, " Applied Energy, 69, pp.293-306
4. Ferdous, D., Dalai, A.K., Bej, S. K., and Thring, R. W.(2002), "Pyrolysis of Lognins: Experimental and Kinetics Studies," Energy Fuel, 16, pp.1405-1412
5. Gomez, C. J., Manya, J. J., Velo, E. and Puigjaner, L.(2004), "Further Applications of a Revisited Summative Model for Kinetics of Biomass Pyrolysis," Ind. Eng. Chem. Res., 43, 901-906
6. Koufopanos, C. Papayannakos, A., N., Maschio, G. and Lucchesi, A.(1991), "Modelling of the Pyrolysis of Biomass Particles. Studies on Kinetics, Thermal and Heat Transfer Effects," Can. J. Chem. Eng., 69, pp.907-915
7. Muller-Hagedornm, M., Bockhorn, H., Krebs, L. and Muller, U.(2003), "A comparative kinetic study on the pyrolysis of three different wood species," J. Anal. Appl. Pyrolysis, 68-69, pp. 231-249
8. Sasauchi, K.(2005), "Application for Electric Generator by Pyrolysis Gasification of Biomass," J. Com. Soc. Japan, 47-139, pp.31-39(in Japanese)
9. Tanoue, K., Nishimura, T., Hinauchi, T., Uchida, K., Taniguchi, M. and Sasauchi, K.(2007), "Analysis of pyrolysis of woody biomass with heating rate," Energy and Resources, 28-3, pp.186-192 (in Japanese)
10. Ueoka, K., Matsushita, Y., Yamamoto, T., Aoki, H., Miura, T.(2004), "Experimental Investigation of Biomass Pyrolysis Mechanism Using Cellulose as a Model Compound," JASME (B), 70-694, pp.1598-1603 (in Japanese)
11. Vamvuka, D., Kakras, E., Kastanaki, E. and Goammelis, P.(2003), "Pyrolysis characteristics and kinetics of biomass residuals mixtures with lignite," Fuel, 82, pp.1949-1960
12. Varhegyi, G., Jakab, E., Antal, M. J.(1994), "Is the Broido-Shafizadeh model for cellulose pyrolysis true?, " Energy and Fuels, 8, pp.1345-1352
13. Willner, T. and Brunner, G.(2005), "Pyrolysis Kinetics of Wood and Wood Component," Chem. Eng. Technol., 28-10, pp.1212-1225

498e Reactor Characteristics for the Suspended-Catalyst Scwg Process

Yukihiko Matsumura[1], Shotaro Hara[2], Shuhei Inoue[2], Tomoaki Minowa[3], Yoji Noda[4], and Yoshihisa Shimizu[5]. (1) Hiroshima University, 1-4-1 Kagamiyama, Higashi-hiroshima-shi, Hiroshima, 739-8527, Japan, (2) Department of Mechanical System Engineering, Hiroshima University, 1-4-1 Kagamiyama, Higashi-hiroshima-shi, Higashi-Hiroshima, 739-8527, Japan, (3) Biomass Technology Research Center, National institute of Advanced Industrial Science and Technology (AIST), 2-2-2 Hiro-Suehiro, Kure, 737-0197, Japan, (4) Toyo Koatsu Co., Ltd., Hiroshima, Japan, (5) The Chugoku Electric Power Co., Inc., Hiroshima, Japan

The supercritical water gasification of chicken manure was studied using suspended activated carbon catalyst. Employing suspended catalyst is a new idea so that reactor plugging is to be avoided in supercritical water gasification reactor. The feedstock used in this study is the egg layer's manure, and collected from the farm. It was first pulverized with a commercial continuous pulverizer to particle size less than 500 micrometer, and then hydrothermally pretreated. Then the activated carbon powders made from coconut shell was added. The reactor employed was a laboratory-scale continuous reactor with inner diameter of 2.16 mm and length of 12 m. The feedstock was fed to the reactor with a piston, heated to the reactiont temperature in a preheater, and then gasified in the reactor. To determine the reaction characteristics, the concentration of the feedstock itself and the amount of added catalyst was changed. The residence time of the feedstock in the reactor was also changed. The reaction temperature was another parameter. The activated carbon catalyst could be recovered at the in-line liquid-solid separator, and the resulting effluent was clear water. The product gas was separated from the liquid phase after depressurization, and its volumetric flow rate was determined with a wet gas meter. The product gas composition was measured with gas chromatograph. The liquid phase was analyzed with TOC. From these anlaysis, carbon gasification efficiency was determined for each conditions. Chemical reaction engineering analysis on the resulting data was conducted to determine the chemical reaction characteristics for this proposed catalyst-suspended supercritical water gasification reactor.

498f Biocarbon Production From Banagrass Using Flash Carbonization™ Process For Metallurgical Use

Takuya Yoshida[1], Russell Yost[2], and Michael J. Antal Jr.[1]. (1) HNEI, University of Hawaii at Manoa, 1680 East-West Road, POST 109, Holunolu, HI 96822, (2) Department of Tropical Plant and Soil Sciences, University of Hawaii at Manoa, 3190 Maile Way, St. John 102, Honolulu, HI 96822

The Flash Carbonization™ (FC) process, which was developed in University of Hawaii, can carbonize biomass feed in less than 50 min of reaction time and can produce large quantities of inexpensive biocarbon. Its fixed-carbon yield can reach the thermochemical equilibrium limit. In this process the biomass feed is carbonized mainly by self-sustained combustion with downward air flow. Consequently the process conditions such as air flow rate and air delivery amount have significant effects on the reaction and on the characteristics of product charcoal. Moreover the carbonization reaction can vary depending on biomass species. Banagrass, a variety of elephantgrass (*Pennisetum purpureum*), has high growing rate and appears to be suitable for tropical areas such as Hawaii, which means that banagrass can be a good feed for the production of biocarbon. However the thin structure of herbaceous species like banagrass could make it difficult to get high fixed-carbon yield. As for utilizations of biocarbon from banagrass, use as a metallurgical reductant is attractive especially in tropical countries like Brazil, where eucalyptus charcoal has been used in metallurgical industry. In this work, banagrasses with different maturities have been carbonized using FC process and the effects of the process conditions and the maturity of banagrass on the yields and the characteristics of charcoal are discussed. Feasibility of utilization of banagrass biocarbon in metallurgy is also discussed.

A Novel Spouted Bed Bioreactor for Solid State Fermentation of Value Added Chemicals and Enzymes

Patrick Bennett and Shang-Tian Yang

The bioprocessing industry has sought efficient bioreactor systems capable of utilizing low value commodities or byproducts from the U.S. agricultural industry for the production of food ingredients, chemicals, and biologically active compounds. Solid State Fermentation (SSF) is capable of employing abundant, low-cost biomaterials with little or no pretreatment to efficiently produce high valued products. SSF has unique process characteristics that provide advantages over submerged culture fermentation. There is little or no free water in the fermentation system allowing for high volumetric product concentration and easy downstream processing. The major challenge to SSF is difficulty in solids handling and hold up due to aggregate formation from mycelial caking. A gas-solid spouted bed bioreactor (SBB) for SFF is proposed. The SBB is capable of preventing agglomeration of particles because high speed impacts occur in the core region of the spouted bed. It is also capable of handling the large coarse particles which are to be used as substrates for the SSF process. Additionally, the spouted bed bioreactor provides exceptional mixing which is advantageous for mass and heat transfer. The efficacy of solid state fermentation to produce important enzymes and industrial chemicals will be demonstrated by fermentation of *Trichoderma reesei* to produce cellulase. The fermentation conditions will then be optimized with respect to substrate, initial moisture content, temperature, fermentation time, inorganic salt concentration and surfactant concentration. After optimal conditions are ascertained, growth kinetics will be determined for use in a mathematical model which will simulate the spouted bed bioreactor. The model will be validated through experimentation and used to design a pilot scale reactor.

A Sustainable Synthesis of Betaine Amphoteric Surfactant Through α-Chloro-Fatty Acid

Wang Ze-yun[1,2,3], Xue-yi Hu [2], *Fang Yun*[2♣], *Zhang Shu-fen*[1♣♣]

(1. State Key Laboratory of Fine Chemicals, Dalian University of Technology, Dalian 116012, China;

2. School of Chemical and Material Engineering, Southern Yangtze- Nafine Center, Southern Yangtze University, Wuxi, Jiangsu 214122, China;

3. Nafine Chemical Industry Group Co., Ltd., Yuncheng, Shanxi 044000, China,)

Abstract: Betaine like surfactant, such as N-substituted betaine, has been synthesized with long chain alkyl amine and *α*-chloroacetic acid as starting materials. A sustainable synthesis of an amphoteric C-substituted betaine surfactant, *α*-capryl betaine (*α*-CB), derivatized from *α*-chlorofatty acid is described in this paper. The reaction conditions were optimized with orthogonal test. The yield of *α*-CB reached 88.3% in 10h at 85°C with the initial molar ratio of trimethylamine to *α*-chlorododecanoic acid of 3:1.

Keywords: dodecanoic acid; betaine; chlorination; detergent; surfactant; synthesis; personal care products

1. Introduction

Betaine like surfactants such as N-substituted alkyl betaine (commercial product BS12) and cocamidopropyl betaine (commercial product CAPB) are of excellent amphoteric surfactants and the essential secondary surfactants in shampoos, body care products and toiletries (Zhang *et al.*,2001). For decades, they have been the popular ingredient employed in personal care products as a gentle detergent enhancing mildness, improving foam and softening water. So far, betaine like surfactants have been synthesized using *α*-chloroacetic acid and long chain alkyl tertiary amine(scheme 1) or fatty amidopropyl dimethylamine as the starting materials.

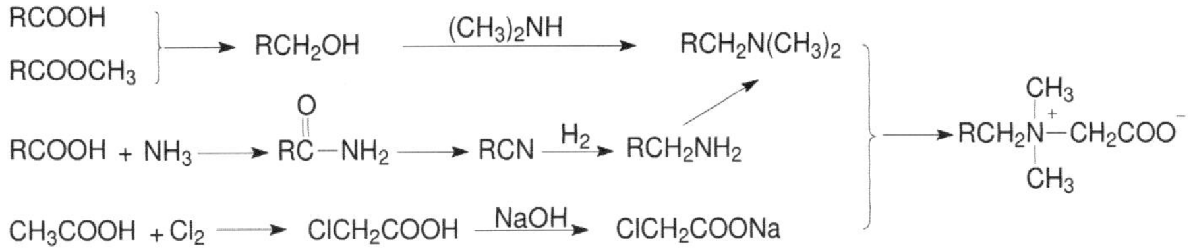

Scheme 1 The synthesis route of N-substituted alkyl betaine

Unfortunately, the remains of *α*-chloroacetic acid or amidoamine in the product causes

Correspondence to:
♣ Tel.86-510-89880600, Fax: 86-510-85917920; E-mail: yunfang@126.com
♣♣ Tel.86-411-88993621; E-mail: zhangshf@chem.dlut.edu.cn

irradiation to skin, which is almost the only concern left in the application of alkyl betaine surfactant (Fowler *et al.,* 1997). However, betaine like surfactants still draw much attention (Danov *et al.*, 2004; Delgado *et al.*, 2006; Li *et al.*, 2005; Ono *et al.*, 2005) because people could hardly find alternatives which can beat them with these excellent surface activities. Much affords have been made on looking for new structure amphoteric surfactants (Rasika *et al.*, 2006; Wang *et al.*, 2002).

α-Alkyl betaine is the alternative of BS12 or CAPB, which can be obtained from α-fatty acid and trimethylamine. Frederlck (Fredrlck *et al.,*1938) and Fang *et al.* (Fang *et al.*, 1993; Xia *et al.*,1994) attempted using α-boro-fatty acid and trimethylamine to synthesize α-alkyl betaine, but little luck was found with α-chloro-fatty acid with the same synthesis route(Fang *et al.*, 1994).

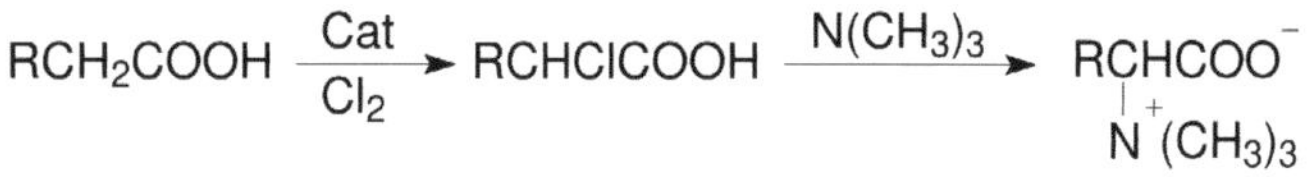

Scheme 2 The synthesis route of C-substituted alkyl betaine

In this study, starting from α-chloro-fatty acid, aiming at synthesizing α-alkyl betaine, we firstly investigated the single factor affected the synthesis, then optimized the reaction parameters with orthogonal test, and tested the surface activity of the product.

2. Experimental

2.1 Chemicals

α-chlorododecyl acid (97% pure by GC), made from dodecyl acid and Cl_2, synthesized and purified according to reference(Cheng *et al.*, 1992; Paatero *et al.*, 1992; Salmi *et al.*, 1993; Xia *et al.*, 1994); Trimethylamine (33%,aqueous), CP, from Shanghai Chemicals Co, China. All chemicals were used as received.

2.2 Synthesis and Analysis

To an autoclave equipped with mechanical stir, heating and temperature controlling system, a mixture of α-chlorododecyl acid, trimethylamine and water was added and stirred in desired hours at desired temperature, and then cooling down to the room temperature before decant the reaction mixture. The content of α-decyl betaine and its yield were measured and calculated via HPLC analysis.

The reaction mixture was adjusted to pH11 with NaOH aqueous solution, the unreacted trimethylamine was then removed by rotary evaporation under vacuum. The residue was adjusted to pH2 with hydrochloride and then dehydrated the product. The produced white powder was solved in ethanol and filtered to remove NaCl, dried and recrystalized from acetone, finally a white crystal was obtained for the subsequent testing.

Surface tension was tested by using drop volume method.

Foamability testing was carried out by measuring the height of foam formed from a 0.1% product solution (in 250ppm calcium hardness water) using Ross-Miles test at 40°C. The initial foam height was measured right after the solution had flowed. The foam height after 5min was determined to indicate the foam stability.

The ability of the product to soften water as a lime soap dispersing agent is defined as lime soap dispersing ability (LSDA) and measured according to ISO 6387-1983, an acidimetric method (Modified Schoenfeldt method).

3. Result and discussion

3.1 Single factor experiment on reaction parameters

We firstly investigated the effect of reaction temperature on the yield of the betaine; meanwhile, since the aminolysis of chlorocarboxylic acid is a nucleophilic displacement and the reaction mixture generated lots of fine foams which might influence the mass transportation, so we also tested the addition of cthanol into thc rcactants mixture to alternate the solvent polarity and defoam, and of cause, it also help to dissolve α-chlorododecyl acid in water.

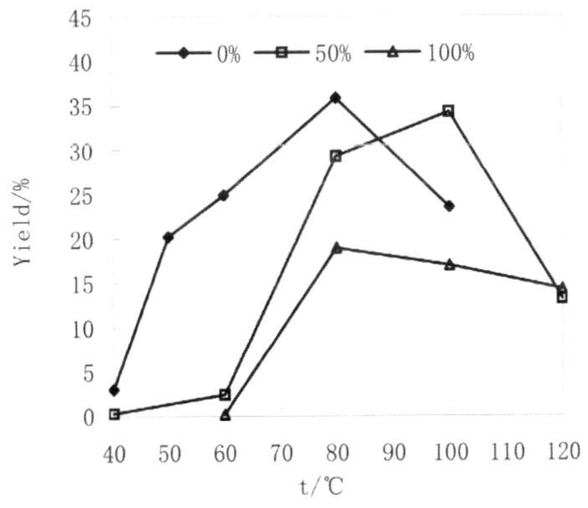

Figure1 Effect of temperature on the reaction with different ethanol dosage.

10h, α-chlorododecyl acid, 16.7g/100mL ; trimethylamine: α-chlorododecyl acid=3:1(mol/mol); ethanol content in the reaction mixture: ■0, □50%, △100%

Figure1 indicates that adding ethanol did not help any in improving the yield, even worse; and the optimum reaction temperature is 80°C.

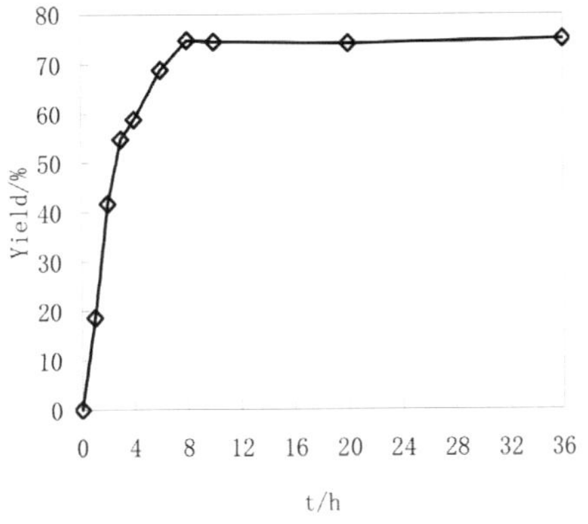

Figure2 Time course of the reaction.

α-chlorododecyl acid, 26.6g/100mL; trimethylamine: α-chlorododecyl acid =3:1(mol/mol); 80℃

We then increased the concentration of α-chlorododecyl acid to 26.6g/100mL, the consequential reaction time course is shown as figure2. So the reaction reached equilibrium in 8-10hrs with a higher yield of the betaine.

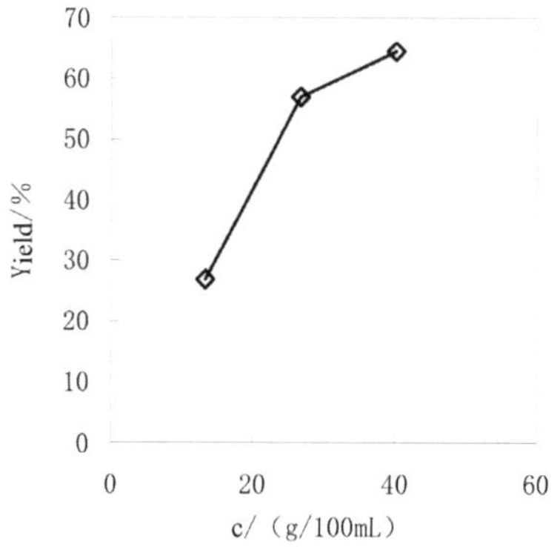

Figure3 Effect of concentration of α-chlorododecyl acid on the reaction.

10h; trimethylamine: α-chlorododecyl acid =3:1(mol/mol); 80℃

Although further raising the initial concentration of α-chlorododecyl acid would increase the yield, but there is a limitation due to the solubility of both α-chlorododecyl acid and trimethylamine. For this reaction, the temperature, the initial concentration of the starting materials, these factors are interrelated; therefore, we made an orthogonal test to optimum the reaction in the following section.

3.2 Orthogonal test to optimize the reaction condition

Table 1 Orthogonal test for optimizing the reaction parameters

Run	Reaction time /h	Reaction temperature /°C	concentration of α-chlorododecyl acid / (g/100mL)	trimethylamine: α-chlorododecyl acid /(mol:mol)	Yield of betaine /%
1	5	60	13.3	1:2	2.34
2	5	80	26.7	1:3	62.08
3	5	100	40.0	1:4	81.70
4	10	60	26.7	1:4	45.94
5	10	80	40.0	1:2	68.25
6	10	100	13.3	1:3	39.55
7	15	60	40.0	1:3	37.10
8	15	80	13.3	1:4	56.04
9	15	100	26.7	1:2	39.50

The data from orthogonal experiment were processed with range analysis, variance analysis, correlation analysis and regression analysis. The tracing range analysis was managed to find the main factors affecting the yield. It was determined that the concentration of α-chlorododecyl acid was the most significant factor. And with level analysis and the consideration of the process flexibility, we found the optimum reaction conditions are: 10h, 80℃, 40g/100mL of α-chlorododecyl acid, trimethylamine: α-chlorododecyl acid=3:1(mol/mol); with which a yield of 88.4% was obtained.

3.3 Surface activities of α-capryl betaine

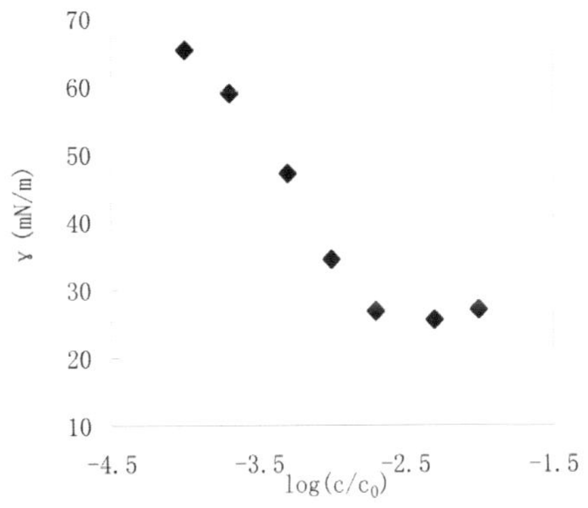

Figure 4 γ-log c curve of the aqueous solution of α-capryl betaine.

From figure4, the cmc of α-capryl betaine is 1.7×10^{-3}mol/L, and γ_{cmc} is 25.2mN/m. Except the surface tension, we tested and compared the foam ability and lime soap dispersing power of this alternated C-substituted betaine surfactant and the conventional N- substituted betaine surfactant, BS12, as shown in Table2.

Table 2 Foamability and lime soap dispersing power of α-CB and BS-12

	α-CB		BS-12	
Foamability(cm)	0min	5min	0min	5min
	17.9	15.6	17.3	14.8
LSDA (%)	19.90		83.01	

Clearly, α-CB has similar foamability with that of BS-12, but better lime soap dispersing power.

4. Conclusion

A sustainable synthesis of amphoteric betaine surfactant, C-substituted α-capryl betaine (α-CB) derivatized from α-chlorofatty acid is achieved. The yield of α-CB reached 88.3% in 10h at 85°C with the initial molar ratio of trimethylamine to α-monochloro dodecanoic acid of 3:1, which are the reaction conditions optimized with orthogonal test. The product has similar foamability with that of

BS-12, but better lime soap dispersing power.

Acknowledgements

This work is supported by National Key Technologies Research and Development Program of China during the 11st Five-year Plan (2007BAE52B01), we gratefully thank for the financial support.

Reference

Cheng, B. C.; Xia, P. F.; Peng, W. H., Synthesis of Mid-Long Chain α-Chloroaliphatic Acids. *Chinese Journal of Applied Chemistry*, **1992**, 9, 83-85.

Danov, K. D.; Kralchevska, S.D.; Kralchevsky, P. A., Mixed Solutions of Anionic and Zwitterionic Surfactant (Betaine): Surface-Tension Isotherms, Adsorption, and Relaxation Kinetics. *Langmuir*, **2004**, 20, 5445-5453.

Delgado, C.; Merchan, M. D.; Velazquez, M. M., Effect of Surfactant Structure on the Adsorption of Carboxybetaines at the Air-Water Interface. *Colloids Surf., A.* **2006**,280, 17-22.

Fang, Y.; Xia, Y. M.; Jiang, W. Z.; He J. R., Synthesis and Properties of Zwitterionic Surfactants(I) Synthesis of α-Bromination Long Chain Fatty Acid. *China Surfactant Detergent & Cosmetics.* **1993**, 5, 1-6.

Fowler, JF; Fowler, LM; Hunter, JE., Allergy to cocamidopropyl betaine may be due to amidoamine: a patch test and product use test study. *Contact Dermatitis,* **1997**, 37 (6): 276-281

Fredrlck, B. D.; Carneys, P.; Frank W. J.; Frank, W. J.; Penns G., Nitrogen Containing Organic Compounds. US 2,129,264.**1938**.

Li, YM; Xu, GY; Luan, YX; Yuan, SL; Zhang, ZQ., Studies on the interaction between tetradecyl dimethyl betaine and sodium carboxymethyl cellulose by DPD simulations. *Colloids and Surfaces A-Physicochemical and Engineering Aspects,* **2005**, 257-58: 385-390.

Ono, Y.; Shikata, T., Dielectric Behavior of Aqueous Micellar Solutions of Betaine-Type Surfactants. *J. Phys. Chem. B.* **2005**, 109, 7412-7419.

Paatero, E.; Tapio, S., Krister F., Selective Synthesis of α-Chlorocaarboxylic Acids**,** *Ind. Eng. Chem. Res.***1992**, 31, 3425-3437.

Rasika M. Satam A; Manohar R. Sawant A Synthesis of Amphoteric Surfactants via Esterification Process. *Journal of Dispersion Science and Technology*, **2006**, 27: 407 - 411

Salmi, T.; Paatero E.; Fagerstolt, L., Optimal Degeree of Backmixing in Autocatalytic Reactions, a Case Study: Chlorination of Dodecanoic Acid. *Trans IchemE.* **1993**, 71, 531-542.

Wang, SF; Furuno, T; Cheng, Z., Synthesis of new betaine-type amphoteric surfactants from tall oil fatty acid. *Journal of Wood Science*, **2002**, 48 (5): 419-424.

Xia, Y. M.; Fang, Y.; He, J. R., Synthesis and Properties of zwitterionic surfactant(II) Synthesis and properties of a long chain alkyl betains. *China Surfactant Detergent & Cosmetics.*

1994,4,1-3.

Xia, Y. M.; Fang, Y.; Zhang, M. S., Synthesis and Poperties of Zwitterionic Surfactants(I), Synthesis of α-Chlorization Long Chain Fatty Acid. *China Surfactant Detergent & Cosmetics.* **1994**, (3), 1-3.

Zhang, Z. Q.; Xu, G. Y.; Ye, F., Surface Activity of Mixed System of Dodecyl Betaine and Sodium Dodecyl Sulphate. *Acta Physico-Chimica Sinica*, **2001**, 17, 1122-1125.

513c Cell Wall Deconstruction Of Corn Stalks As A Function Of Hot Water Pretreatment And Enzyme Hydrolysis

Michael Ladisch[1], Meijuan Zeng[2], Debra Sherman[3], Chiang-ping Huang[3], Nathan Mosier[4], and Wilfred Vermerris[5]. (1) Agricultural & Biological Engineering, Biomedical Engineering & Laboratory of Renewable Resources Engineering (LORRE), Purdue University, Potter Engineering Center, 500 Central Drive, West Lafayette, IN 47907-2022, (2) Agricultural and Biological Engineering and Laboratory of Renewable Resources Engineering (LORRE), Purdue University, Potter Engineering Center, 500 Central Drive, West Lafayette, IN 47907-2022, (3) Life Science Microscopy Facility, Purdue University, West Lafayette, IN 47907-2022, (4) Agricultural and Biological Engineering & Laboratory of Renewable Resources Engineering (LORRE), Purdue University, Potter Engineering Center, 500 Central Drive, West Lafayette, IN 47907-2022, (5) Agronomy, University of Florida, Gainesville, Institute of Food and Agricultural Sciences, P.O. box 110500, Gainesville, FL 32611-0500

Hydrolysis experiments and SEM studies performed on different fraction of leaves, stalk fiber, stalk pith from corn show significant differences response to hot water pretreatment where the biomass is cooked in water at 190°C for 15 min. Pith is the most readily enzyme hydrolyzed fraction both before and after hot water pretreatment. After hot water pretreatment, glucose and xylose formation are significantly enhanced and accompanied by dramatic cell wall structural changes. Enzymatic hydrolysis on pretreated pith show that both glucose and xylose conversion can reach 90% in 24 hours at 60FPU/g glucan cellulase loading while pretreated leaves and fiber achieve only 70% and 50% respectively, under the same conditions. Pretreated leaves, fiber and pith have comparable lignin contents, but a 50% difference in hydrolysis of different fractions suggests lignin structure is an important parameter in determining extent of cellulose hydrolysis. The impact of pretreatment and cell wall features on the extent of cellulose hydrolysis is discussed. The utility of microstructural analysis in selecting combinations of pretreatment and enzyme hydrolysis conditions that maximize yields is presented.

513d Bioconversion Of Paper Mill Sludge To Ethanol

Li Kang and Y. Y. Lee. Chemical Engineering, Auburn University, 212 Ross Hall, Auburn University, Auburn, AL 36849

Paper mill sludges are solid waste material generated from paper making process. They are composed of pulp residues and inorganic additives. Because of high carbohydrate content, the pulp mill sludges are potential feedstock for bioconversion into value-added products. The carbohydrate portion of the sludges has similar chemical and physical characteristics as pulp. Therefore the bioconversion may be applied without pretreatment, a significant economic benefit. On the other hand, the sludges also contain high level of inorganics including calcium and aluminum salts, which may adversely affect the bioconversion process. In this study, primary and recycle sludges collected from a paper mill were investigated as feedstock for bioconversion to ethanol. The sludges were subjected to standard enzymatic digestion test with and without pH adjustment. Enzymatic digestion of the sludges without pH adjustment yielded very low sugars. After treatment with sulfuric acid, the sludges were easily digested to give 70-80% of sugar yield. The feedstocks were also put through simultaneous saccharification and Co-Fermentation (SSCF) using Spezyme-CP and recombinant E. coli (KO11). The yield of ethanol from the SSCF operated without pH control exceeded the yield of ethanol calculated from the enzymatic digestion data. The ethanol yields based on total glucan and xylan were in the range of 85-90% of theoretical maximum. The SSCF proceeded well without pH control whereas the enzymatic digestion required pH control. It was due to the fact that the sugars generated by enzyme hydrolysis were adsorbed on the salts in the sludge, most of which is calcium carbonate. In the SSCF, sugars released by cellulase enzyme were rapidly consumed by microorganism, which prevented adsorption of sugars.

Pretreatment and Enzymatic Hydrolysis of Soybean Hulls[a]

D.Y. Corredor[b], X.S. Sun[c], J.M. Salazar[d], K.L. Hohn[d], and D. Wang[b,e]

ABSTRACT

Soybean hull is a potential feedstock for production of bio-ethanol due to its high carbohydrate content (≈50%) of nearly 37% cellulose. Soybean hulls could be the ideal feedstock for fuel ethanol production, because they are abundant and require no special harvesting and additional transportation costs as they are already in the plant. The objective of this research was to study the performance of dilute acid and steam-explosion pretreatment technology on degradation of soybean hulls, and to identify the optimum pretreatment and enzymatic hydrolysis conditions for converting soybean hulls into fermentable sugars. The effects of reaction time, temperature, acid concentration and type of acid on hydrolysis of hemicellulose in soybean hull and total sugars yield were studied. The combination of acid (H_2SO_4, 2% w/v) and steam (140 °C, 30min) efficiently solubilized the hemicellulose, giving a pentoses yield of 96%, and hexose yield (23 mg/ml) corresponding to 72% of the theoretical. Maximum total sugars yield, achieved with a combination of treatment and enzymatic hydrolysis, was 80% of the theoretical.

Keywords: pretreatment, enzymatic hydrolysis, soybean hull, dilute acid, steam explosion

INTRODUCTION

Ethanol derived from biomass such as grains, grain residues, agricultural by-products, and dedicated energy crops has great potential to be a sustainable replacement of transportation fuels and other potential applications. A dramatic increase in ethanol production using the

---

[a] Contribution from Kansas State Agricultural Experiment Station, Manhattan, KS 66506
[b] Department of Biological and Agricultural Engineering, Kansas State University, Manhattan, KS 66506
[c] Department of Grain Science and Industry, Kansas State University, Manhattan, KS 66502
[d] Department of Chemical Engineering, Kansas State University, Manhattan, KS 66506
[e] Corresponding author. Phone: 785-532-2919. Fax: 785-532-5825. Email: dwang@ksu.edu

current grain-starch-based technology may have resource limitations because grain production for ethanol will compete for the limited agricultural land for food and feed production[1]. Conversion of cellulosic biomass such as agricultural residues to fuels and chemicals offers major economic, environmental, and strategic benefits, and biological processing based on cellulases offers high sugar yields vital to economic success.

The overall amount of monosaccharides accounted for 63%. Soybean hull was also reported to contain 46% cellulose, 18% hemicellulose, and 2 % lignin on a DM basis[2]. Hemicellulose and cellulose content in our sample were in the mid level of the reported range.

About 1.8 billion bushels of soybeans are crushed for oil, protein, and soy flour production in the U.S. This will generate about 10.8 billon pounds of soybean hulls. Monomers of soy carbohydrates from soybean hulls are sugar, which can be used as substrates for bioconversion to produce chemicals and bio-fuels. Mullin and Xiu reported that soybean hulls contain about 16.7% water-soluble substrates and 83.3% water un-extractable substrates[3]. In addition to glucose, soybean hulls also contain xylose, arabinose, galactose, mannose, and uronic acid[4]. Development of an effective method to produce fuel ethanol from soybean hulls will help sustainable economic development and have a great impact on industrial utilization of soybean hulls for value-added products.

Production of chemicals and biobased products from renewable biomass faces significant technical challenges. Success of using biomass for biofuel production depends largely upon physical and chemical properties of the biomass, pretreatment methods, efficient microorganisms, and optimization of processing conditions. The purpose of the pretreatment is to break the lignin seal, pre-hydrolyze the hemicellulose, and disrupt the crystalline structure of the cellulose, enhancing the hydrolysis of cellulose by cellulase.

Mention of firm names or trade products does not constitute endorsement by the U.S. Department of Agriculture over others not mentioned.

Enayati et al. and Ouhida et al., studied enzymatic saccharification of soybean hull-based materials and demonstrated ease of biodegradability of soy hulls[5,6]. At present, there is no reported information on pretreatment, enzymatic saccharification, and fermentation of soybean hulls for biofuel. Although several methods are available for biomass pretreatment, selectivity of pretreatment methods is highly restricted by the nature of the raw materials. Objectives of this research were to study the performance of dilute acid and steam explosion pretreatments on degradation of soybean hull and to identify optimum pretreatment and enzymatic hydrolysis conditions for converting soybean hulls into fermentable sugars.

MATERIALS and METHODS

Materials

Soybean hulls with 12% moisture content and particle size <1 mm were obtained from the Cargill Company (Dayton, Ohio), and were stored at 4 °C. The chemical composition of soybean hulls is about 36% cellulose, 12% hemicellulose, 2% starch, 12% acid-soluble lignin, and 14% protein (Table 1). Total carbohydrate composition was about 50%. Theoretical yield of sugars from soybean hulls was calculated from the chemical composition as 0.54 g. sugars/g. hull. This transformed into gallons of EtOH would be 84 gallons EtOH/ton dry soybean hulls. Cellulase C2605 from Sigma-Aldrich (102 FPU/ml), Cell-Wall Degrading Complex (Viscozyme L –V2010) (hemicellulase, cellulase, arabinase and xylanase complex), and Novozyme 188 (β-glucosidase) (250 CBU/ml) from Novozyme (U.S. Office: Franklinton, N.C.) were used for enzymatic hydrolysis of soybean hulls into fermentable sugars. Sugars used for HPLC calibration were purchased from Fischer Scientific Inc.

Enzymatic Hydrolysis

Three commercial enzymes (cellulase C2605, Novozyme 188, and Cell-Wall Degrading Complex V2010) were used for hydrolysis of cellulose in soybean hulls. Effects of a combination of cellulase and β-glucosidase at enzyme loading of 15 FPU/g. cellulose and 50 CBU/g. cellulose, respectively, with cell-wall degrading complex added at a ratio 1:1 (v/v) cellulase/hemicellulase on hydrolysis of soybean halls were studied. Enzymatic hydrolysis was carried out in flasks with 100 ml of slurry at 45°C and pH 4.8 for 12 to 96 h in a water-bath shaker with agitation speed of 140 rpm.

Pretreatment with Dilute Acid and Steam Explosion

The treatment was carried out in a 1-L pressure reactor apparatus (Parr Instrument Company). Soybean hulls were mixed with dilute acid to obtain 10% dry matter. The slurry was loaded into the reactor and treated at the desired temperature and time. After treatment, the slurry was removed from the reactor by releasing the pressure through the liquid valve to achieve "explosion"**Error! Bookmark not defined.**. To reach the desired temperature in a minute, 90 psig saturated steam was injected.

Analytical Methods

Cellulose and hemicellulose of soybean hulls were analyzed by Filter Bag Technology (ANKOM Technology, Macedon, N.Y.). Acid-soluble lignin was determined by laboratory procedures, which were developed by the National Renewable Energy Laboratory (NREL)[7]. Concentrations of sugars were determined by HPLC using an RCM-monosaccharide column (300 x 7.8 mm; Bio-Rad, Richmond, Calif.) and refractive index detector.

Soybean hulls before and after pretreatments were analyzed by XRD in a Bruker AXS D-8 difractometer settled at 40 KW, 40 mA; radiation was cupper Kα (λ= 1.54 Å), grade range between 5 to 40° with step size of 0.03°. Images of the surface of pretreated and untreated soybean hulls were taken at magnifications from 1.5 K to 3K using a Hitachi S-3500 N Microscope scanning electron microscopy (SEM).

RESULTS AND DISCUSSIONS

Enzymatic Hydrolysis of Soybean Hulls

Sugars yield from enzymatic hydrolysis (EH) using cellulase, β-glucosidase, and hemicellulase complex without pretreatment is shown in Figure 1. Sugars yield increased when both cell-wall degrading complex (hemicellulase, cellulase, arabinase and xylanase complex) and cellulase enzymes were used for enzymatic hydrolysis. The highest sugar yield was 27.5%, obtained after 36 h of enzymatic hydrolysis. The addition of the cell-wall degrading complex increased the sugar yield by more than 97%, compared with the enzymatic hydrolysis using cellulase alone. Both enzyme concentration and hydrolysis time had a significant effect on sugars yield.

Pretreatment with Hydrochloric Acid and Steam Explosion

Soybean hulls treated with HCl and steam explosion gave a maximum pentose yield (75%) obtained at 140°C for 30 min and 1% HCl (Table 2). Total sugar yield based on total amount of sugars (0.54 g. Sugars/g. hulls) was only 19.7%, because hexoses were not recovered from this pretreatment. Pentose yield increased as both reaction time and temperature increased. This suggests that longer time with higher temperature is useful to hydrolyze and degrade the hemicellulose layer. It was demonstrated that thermal expansion coefficients of hemicellulose in aqueous solutions were several times greater than that of hemicellulose in the dry samples[8] . The high recovery of pentose obtained using this pretreatment was also probably due to the high temperature reached by steam increased thermal expansion of the hulls and made the hydrogen bonds easily rupturable.

Pretreatment with H_2SO_4 and Steam Explosion

Steam explosion with 1% H_2SO_4 at 140°C for 30 min gave a pentose yield of 90%, while using 2% H_2SO_4 at the same conditions gave a maximum pentose yield of 96% (Table 3). Compared with hydrochloric acid and steam explosion, this treatment increases the pentose yield from 74% up to 96% and is the most effective method for hydrolysis of hemicellulose. This suggests that both faster heating with saturated steam and use of a strong acid such as H_2SO_4 are powerful to remove and hydrolyze hemicellulose, and could further increase the enzymatic hydrolysis of cellulose.

Hexose yield obtained after 36 h of enzymatic hydrolysis was 72% (Table 4). Compared with the combination of HCl treatment and enzymatic hydrolysis, the combination of H_2SO_4 treatment with enzymatic hydrolysis increased total sugar yield from 27.5 to 79% in the same hydrolysis time of 36 h (Figure 2). The higher hexose yield also indicated that pretreatment with sulfuric acid and steam explosion is an effective method for improving the enzymatic hydrolysis of soybean hulls using the same enzyme loading, and allows for efficient hydrolysis even in the absence of hemicellulase complex enzymes.

Morphological Structure

Morphological features of the soybean hull before and after selected treatments are shown in Figure 3. An SEM image of the untreated soybean hull shows a layer covering the surface of the material and is attributed to hemicellulose and lignin (Figure 3A). The surface layer is also observed in other lignocellulosic materials such as corn stover and sorghum leaves and stems[9]. SEM image of the sample after enzymatic hydrolysis using cellulase and cell-wall degrading complex shows that the compacted outer layer of the soybean hull was partially removed (Figure 3B). SEM image of soybean hulls treated with HCl and steam explosion shows some well-defined micro fibrils (6μm of diameter), which might be an evidence that cellulose particles are agglomerates of individuals cellulose micro fibrils (Fig. 3C). Outer layers seemed to be hydrolyzed or degraded, exposing micro fibrils of cellulose to enzymatic attack. These fibrils still appeared to be entangled by some amorphous material, probably un-removed hemicellulose. Figure 4D shows single cells after treatment with sulfuric acid and steam explosion, resulting in fibers that have a relatively clean and smooth surface. The micro fibrils were also separated from the initial connected structure and fully exposed, thus increasing the external surface area and porosity.

These fibrils could be evidence of the presence of micro-crystalline cellulose fibrils exposed in the remaining solid after pretreatment, suggesting that pretreatment is critical to expose cellulose to enzymatic hydrolysis. To validate this, soy hulls and solids after pretreatment were analyzed by X-ray diffraction (XRD). Figure 5 shows the XRD spectra of the original sample, pretreated sample with sulfuric acid and steam explosion, and the remaining solids after pretreatment and enzymatic hydrolysis. The XRD of the original sample did not show any crystalline structure in the XRD, probably due to the presence of a thicker amorphous hemicellulose layer covering the cellulose. After pretreatment, the main peaks relative to planes 002 and 020 may be easily observed, showing that cellulose amount increased due to the removal of lignin and hemicellulose. The profile of the diffractogram is in agreement with earlier ones reported in the literature for mycro-crystalline cellulose samples[10,11]. This suggests that pretreatment is effective in exposing cellulose to enzymatic attack.

CONCLUSION

Soybean hulls are a potential feedstock for production of bio-ethanol because of their high carbohydrate content (≈50%) with about 37% cellulose. Pretreatment of soybean hulls could substantially improve recovery of sugars. Overall, results showed that up to 80% of total sugars in soybean hulls were recovered using pretreatment with 2% sulfuric acid and steam explosion at 140°C for 30 min, followed by enzymatic hydrolysis with cellulase (15 FPU/g. cellulose) and β-glucosidase (50 CBU/g. cellulose). This yield is much higher (>100%) than overall total sugars yield obtained by direct enzymatic hydrolysis using not only cellulase and β-glucosidase, but also hemicellulase enzymes with the same amount of enzyme loading and enzymatic hydrolysis conditions. The highest hexose yield of 72% was achieved with a combination of 2% sulfuric acid, steam explosion, and enzymatic hydrolysis.

Table I. Chemical composition of soybean hulls [a].

Component	Soybean Hulls (% db)
Carbohydrates	50.7
Starch	1.75 ± 0.08
Cellulose	36.43 ± 0.09
Hemicellulose	12.48 ± 0.32
Acid soluble lignin	12.21 ± 0.9
Crude fat	3.20 ± 0.4
Crude fiber	32.30 ± 0.32
Crude protein	14.21 ± 0.10
Pectin	6.3 ± 0.2
Ash	4.24 ± 0.03

[a.]Means of four replicates

Table II. Yield (%) of sugars from soybean hulls after hydrochloric acid and steam explosion treatment [a].

Temperature (°C)	Time (min)	Pentoses (mg/ml)	Yield [b] (%)	
			Pentoses	Total
120	5	4.07	28.96	7.62c
	30	8.74	61.55	16.18b
140	5	9.82	69.18	18.19b
	30	10.65	74.97	19.72a

a. The concentration of acid is 1% HCl.

b. Values in the same column with the same letters are not statistically different at $p<0.05$.

Table III. Yield (%) of sugars from soybean hulls after sulfuric acid and steam explosion treatment [a].

Acid (%w/v)	Pentoses (mg/ml)	Yield [b] (%)	
		Pentoses	Total
1.0	12.85	90.50	23.80b
2.0	13.79	96.91	27.55a

a. The reaction temperature was 140°C and reaction time was 30 min.

b. Values in the same column with the same letters are not statistically different at $p<0.05$.

Table IV. Yield (%) of sugars from soybean hulls after sulfuric acid, steam explosion treatment and enzymatic hydrolysis [a].

Hydrolysis Time(h)	Hexoses (mg/ml)	Yield (%)[b]	
		Hexoses	Total
12	21.47	53.10	64.50c
24	28.39	70.21	77.16b
36	29.45	72.82	79.09a

[a.] The pretreatment conditions are 140° C for 30 min, with 2% of H_2SO_4.

[b.] Values in the same column with the same letters are not statistically different at p-value<0.05.

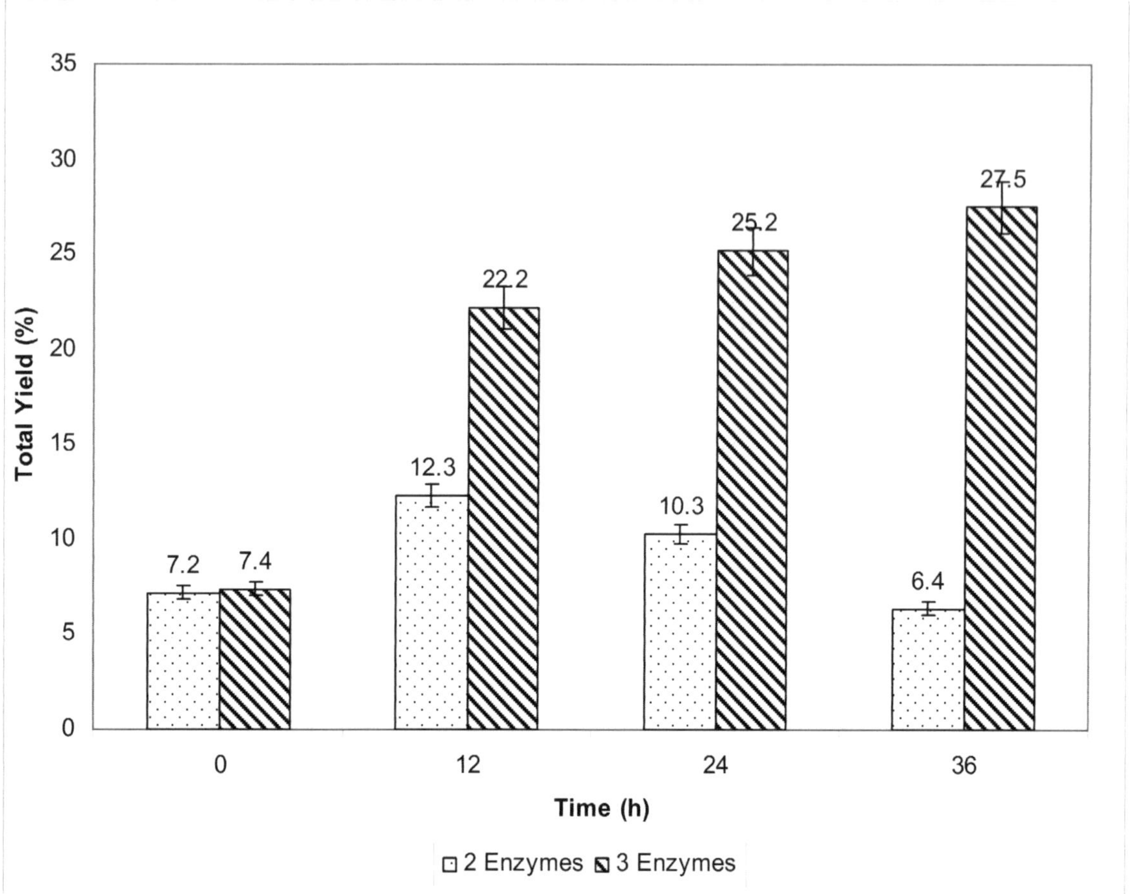

Figure 1. Effects of number of enzymes on sugars yields from untreated soybean hulls: two enzymes (cellulase and β-glucosidase) and three enzymes (hemicellulase , cellulase and β-glucosidase) with an enzyme loading of cellulase 15 FPU/g.cellulose and 50 CBU/g.cellulose of β-glucosidase.

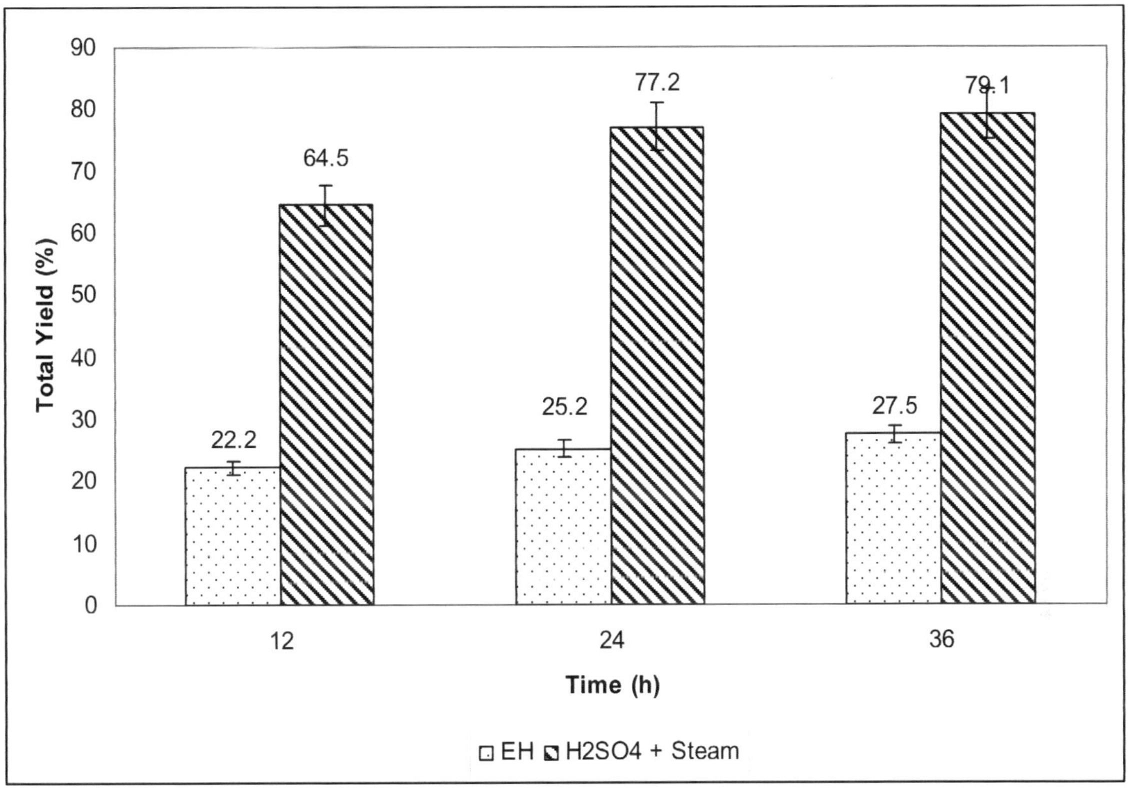

Figure 2. Comparison of total sugars yield (%) from enzymatic hydrolysis (3 enzymes) and combination of pretreatment (2% H_2SO_4, T: 140°C. t: 30 min) and enzymatic hydrolysis (2 enzymes).

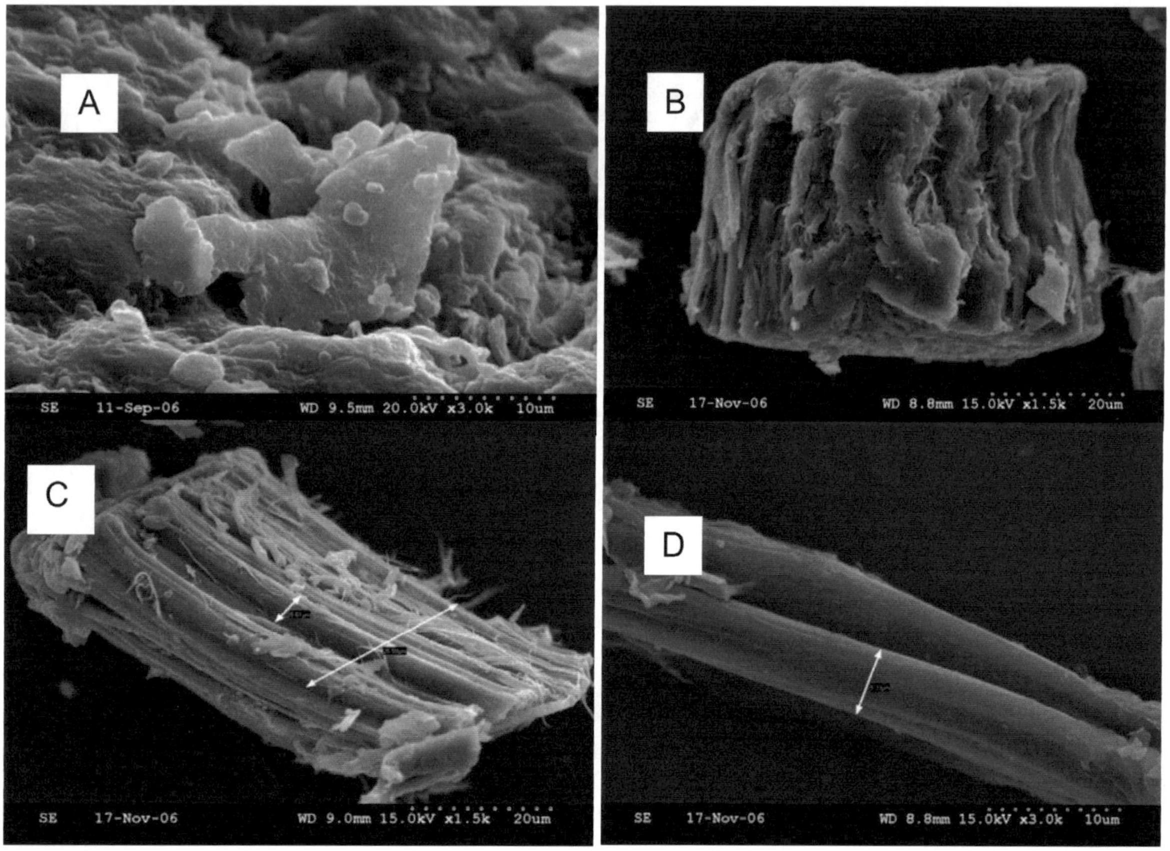

Figure 3. Scanning electro micrograph of a) original soybean hull; b) after enzymatic hydrolysis using cellulase , β-glucosidase, and hemicellulose enzymes for 36 h; c) after treatment with 1% hydrochloric acid and steam explosion (140°C , 30 min); and d) after treatment with 2% sulfuric acid and steam explosion (140°C , 30 min).

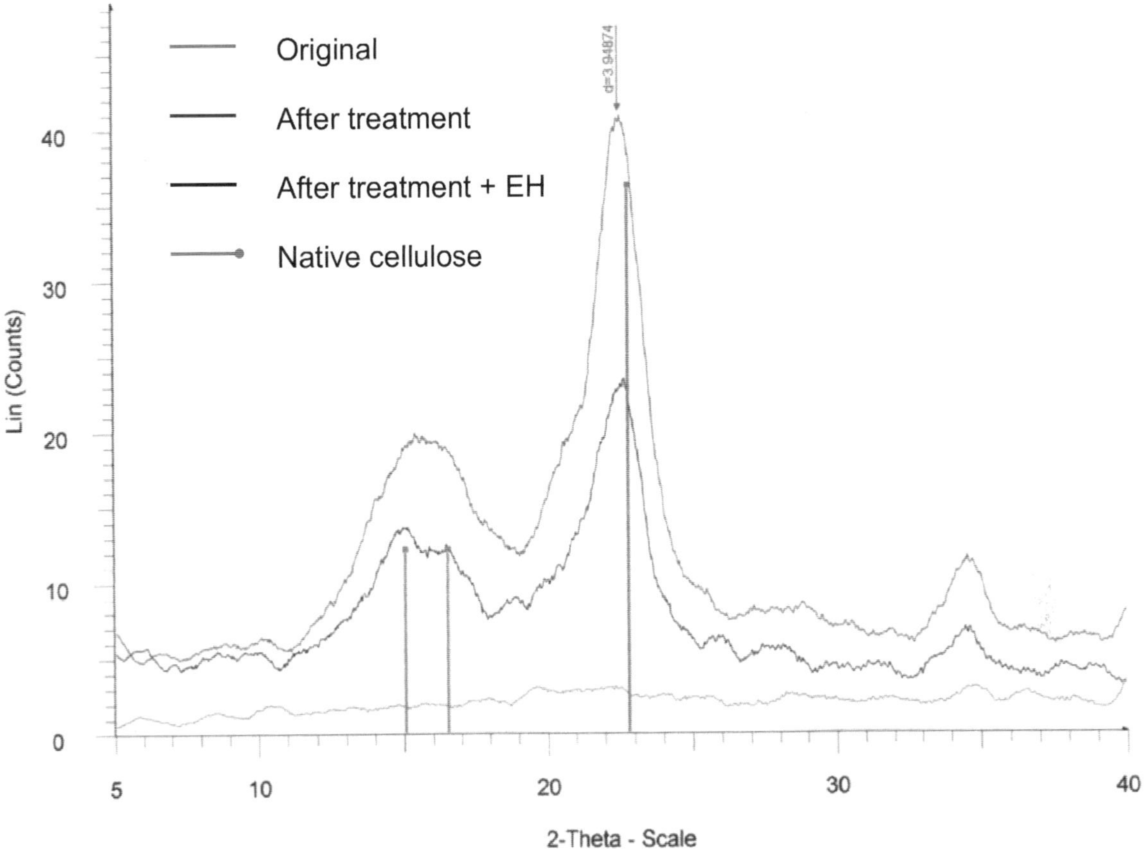

Figure 4. X-ray diffraction of original soybean hull, after treatment with 2% H_2SO_4 at 140 °C for 30 min, and after treatment and enzymatic hydrolysis with cellulase (15 FPU/g.cellulose) and β-glucosidase (50 CBU/g.cellulose). The labeled peak is the principal 002 peak (100% intensity).

LITERATURE CITED

[1] Y. Sun, and J. Cheng. Bioresource Technology 83, 1 (2001).

[2] D.A Blasi, E.C Titgemeyer, J. Drouillard, S. Paisley, and M.J. Brouk, Soybean hulls, composition, and feeding value for beef and dairy Cattle, Kansas State University, Kansas USA (2000).

[3] W.J. Mullin, and W. Xu. J.Agric.Food.Chem. 49, 5331 (2001).

[4] J. Miron, E. Yosel, and D. Ben-Ghedalia. J. Agric. Food Chem. 49(5), 2322. (2001).

[5] N. Enayati and S.J. Parulekar. Biotechnology Progress. 11,708 (1995).

[6] I. Ouhida, J. Perez, and J. Gasa. J. Agric. Food. Chem. 50, 1933 (2002).

[7] NERL Chemical Analysis and Testing Standard Procedures, No. LAP-003-005, National Renewable Energy Laboratory, Golden, CO. (2006).

[8] M.V. Ramiah and D.A.I. Goring. Journal of Polymer Science : Part C. 11, 27 (1965).

[9] T.H Kim and Y.Y. Lee. Bioresource Technology 97, 224 (2006).

[10] S. Ardizzone, F.S. Dioguardi, T. Mussini, P.R. Mussini, S. Rondinini, B. Vercelli, and A. Vertova. Cellulose. 6, 57 (1999) .

[11] J. Soltys, L. Lisowski, and J. Knapczyc. Acta Pharmaceutica Technologica. 30,174 (1984).

LIFE CYCLE ANALYSIS OF BIOMASS PELLETING TECHNOLOGY

Dr. Sudhagar Mani, University of Georgia, Athens, GA

Abstract

Pelletization is a process of transforming loose biomass into dense pellets of specific sizes for easy handling, transport and storage. The process involves with series of operations – drying, size reduction, pelletization (also called densification), cooling, screening and storage/packaging. Drying is the major energy consuming operation, when the biomass is available at high moisture content (~50%). In this study, two types of biomass: sawmill residue (sawdust) and switchgrass as a raw material to produce densified pellets for heat and power generation applications were considered. Sawdust was available for pelleting process at 50% moisture content and required drying steps to produce pellets. Whereas, switchgrass was available at about 15-20% moisture content and did not require any further drying steps before pelletization. A life cycle assessment approach was used to quantify total energy use and environmental emissions for the entire life cycle starting from biomass production, transport to the pellet plant, pelletization process, transport of pellets to the combustion plant and transport of ash to the land application for its final disposal. For sawdust pelleting scenario, energy and environmental emissions to produce sawdust was not considered as it was assumed as a waste material from sawmill (Or allocation of resources were based on the value of the final sawmill products). Whereas for switchgrass pelleting scenario, all the resources used for production of switchgrass was included in the life cycle estimation. Emission inventory data for each of the life cycle stages was used from literature sources, environmental protection agency (EPA) and LCI databases (Franklin USA, Ecoinvent and GHGenius). Environmental impacts of total gaseous emissions were evaluated based on greenhouse gas emission, acid rain formation, smog formation and human toxicity impacts. The energy use and environmental impacts were compared for both sawdust pelleting and switchgrass pelleting scenarios.

513g Cellulosic Ethanol From Afex Treated Oil Seed Cakes And Hulls For Bio-Diesel Processing

Venkatesh Balan, Shishir P. S. Chundawat, Leonardo D. Sousa, Chad A. Rogers, and Bruce E. Dale. Department of Chemical Engineering and Materials Science, Michigan State University, 2527, Engineering Building, East Lansing, MI 48824

Currently, ethanol is made primarily from the corn grain starch in the United States and from cane sugar/ molasses in other part of world. However, grain or cane sugar alone cannot meet the future demands for ethanol and newer feedstocks are being constantly explored. Waste oilseed cakes and hulls are one such suitable alternative feedstock. We propose to manufacture ethanol as a by-product of the vegetable oil refinery that could be integrated for bio-diesel processing. Oil seed cakes and hulls (i.e. rape seed cake, palm seed cake and jatropha seed cake) are made up of cellulosic materials and are waste products left over after extracting the oil.

Through this project we will develop new fundamental understanding of the complex processes occurring when: (1) oil cakes are pretreated by concentrated/aqueous ammonia to enhance the reactivity/ accessibility of structural carbohydrates (cellulose and hemicellulose), (2) pretreated materials are enzymatically hydrolyzed to mixed sugars by synergistic combinations of glucanases, xylanases and other enzymes and (3) these mixed sugars are fermented to ethanol by both native and genetically modified microorganisms. The efficiency of the pretreatment and fermentation process will be evaluated based on the yield of free sugars and ethanol, respectively.

513h Properties and Characterization of Cl-/ZrO2-Fe2O3 for Biomass Liquefaction

Zhiping Le[1], Lizhi Hong[1], Fei Yu[2], Shaobo Deng[2], Roger Ruan[2], Hong Zhang[1], and Zhaoping Chen[1]. (1) Chemistry, Nanchang University, 999 Xuefu Road, Qianhu, Nanchang 330031, China, (2) Department of Biosystems & Agricultural Engineering, University of Minnesota, 1390 Eckles Ave., St. Paul, MN 55108

With the depletion of fossil fuel reserves and the environmental problems, people pay more attention to look for new alternative energy sources. Biomass is abundant in the world, which the best candidate for the renewable sources of energy. In conventional hydrothermal liquefaction process, liquid acid, such as sulfuric acid, was used as the catalyst for the liquid fuels conversion from biomass. This liquid acid could cause serious environmental and post-processing problems, in addition the corrosion of metallic equipments. In this work, a kind of novel catalyst was chosen for the conversion of biomass. The liquefaction of biomass was carried out in a stainless autoclave with rice husk, pubescens and corn stalk as raw materials, ethylene glycol as liquefaction solvent, Cl-/ZrO2-Fe2O3 as catalyst. The results of X-ray Diffraction (XRD) indicted that ZrO2 and Fe2O3 did not form solid solution. ZrO2 was a monoclinic crystal. Fe2O3 had ¨»- crystal phase. The results of liquefaction process showed that the Cl-/ZrO2-Fe2O3 catalyst had high activity for all three biomass used in this experiment. For instance, corn stalk was completely liquefied at 573K, 3h and 4% (wt % of the reactant) Cl-/ZrO2-Fe2O3. The advantage of using solid super acid is that this catalyst was easily separated and can be recycled. After being used five times of Cl-/ZrO2-Fe2O3, the resident of corn stalk in our novel liquefaction process was under 11%. This work is supported by grants from the Jiangxi Provincial Department of Education Fund (High Education & Technology) of China , the Jiangxi Provincial Department of Education Graduate Innovation Fund of China and the Jiangxi Provincial Natural Science Foundation of China (No. 0420032).

513i Cello-Oligosaccharides: A Reaction Intermediate To Understand The Cellulase Reaction Mechanism

Rajesh Gupta and Y. Y. Lee. Chemical Engineering, Auburn University, Auburn, AL 36849

Non-crystalline cellulose (NCC) is a novel product of our laboratory. The crystallinity of NCC is less than 10% and DP is 100-150. Cellulase action on this NCC was found to be very different from crystalline cellulose. Apart from the very high initial hydrolysis rate, NCC also generates cello-olgiosaccharides (COS) as hydrolysis product with cellulase, which are not visible with crystalline cellulose. COS were categorized into two separate fractions: low DP cello-oligosaccharides (LD-COS) and high DP cello-oligosaccharides (HD-COS). LD-COS, from DP 2-7, are detected by HPLC whereas HD-COS are detected only after secondary hydrolysis. On the basis of the profiles of these sugars during enzymatic hydrolysis, individual actions of Exo-glucanase (Exo-G) & Endo-glucanase (Endo-G) are proposed. Other pure cellulosic substrates such as micro-crystalline cellulose (Avicel), á-cellulose, filter paper, cotton are also used in this study. With regard to the reactivity of cellulase, we found that the crystallinity affects the initial rate of cellulose hydrolysis, whereas the DP of cellulose plays an important role in the determination of terminal glucan digestibility. Other findings on the function of individual cellulase components are as follows: (1) Exo-G and Endo-G do not hydrolyze COS. (2) â-G hydrolyzes cellobiose and LD-COS but has negligible reactivity on HD-COS. (3) Reactivity of Exo-G towards the cellulose chain is very low if DP goes below certain value. (4) With NCC, Endo-G mainly works in the beginning of the reaction and produces HD-COS. (5) Reactivity of Endo-G is inversely proportional to the DP of NCC.

513j Ion Exchange Process For The Recovery Of Ionic Liquids

Noureen Faizee, Kripa Rao, Anantharam P Dadi, Sasidhar Varanasi, and Constance A. Schall. Chemical and Environmental Engineering, University of Toledo, 2801 W,Bancroft, University of Toledo, Toledo, OH 43606

Ionic liquids (IL), have gained acceptance as environmentally benign solvents replacing volatile organic compounds as they have virtually no/low vapor pressure. IL's have shown potential as unique designer solvents, the properties of which can be tailored to meet specific application needs by adjusting the structure of thc cations /anions.

Recently IL's that have appreciable cellulose solubility have been discovered[1].We have shown to have enhanced cellulose saccharification yields using pretreatment of cellulose with ionic liquids followed by enzymatic hydrolysis[2].However the economics of commercially using ionic liquids in many industrial processes are still not clear. Current published data do not allow for reliable extrapolation to broad industry application scale. However, it is clear that the ionic liquids which are proposed to be used as solvents need to be recycled and reused in order to have a commercially attractive process.

We would like to present results on the recovery of ionic liquids used for the pretreatment of cellulose and lignocellulosic substrates.This was carried out using solid phase extraction (SPE) using strong cationic exchange resins (SCX) or liquid-liquid extraction.Ionic liquids consist of an organic cation and an inorganic anion (chloride, in this case). The SCX is strongly acidic and exchanges the cation. The ion exchange bed then gets saturated with the cations and can be regenerated by passing hydrochloric acid-a process which allows for recovery of ionic liquid via the reverse ion exchange process.The parameters affecting the regeneration process will be discussed and presented. Also the use of novel ion exchanger configurations such as radial flow ion exchanger membranes will be explored.

References: 1.For,Diego A., Remsin , Richard C.,Swatlosk , Richard P.,Moyn , Patrick , Moyn , Guillermo , Roger , Robin D.Can ionic liquids dissolve wood? Processing and analysis of lignocellulosic materials with 1-n-butyl-3-methylimidazolium chlorid. Green chemistry(2007) volume 9,issue 1(63-69).

2.A. Dadi, P., S. Varanasi, and C.A. Schall, Enhancement of cellulose saccharification kinetics using an ionic liquid pretreatment step.Biotechnology and Bioengineering,95(5) 2006 pg 904-910.

513k Effect Of Ionic Liquid Pretreatment On Fermentation

Kripa Rao, Ananth Dadi, Constance Schall, and Sasidhar Varanasi. Department of Chemical & Environmental Engineering, The University of Toledo, 2801 West Bancroft Street, Toledo, OH 43606

Ionic liquids have been demonstrated to dissolve bio polymers and polysaccharides and use of ionic liquids as effective solvents for cellulose hydrolysis was proposed due to their unique solvating properties and non corrosive nature [1, 2]. This concept was extended for pretreatment of lignocellulosic substrates like corn stover and poplar to give theoretical yields of glucan after enzymatic hydrolysis. Different pretreatment techniques generate product streams which behave differently in the downstream processes. The fermentation of the hydrolysate obtained from any pretreatment processes is directly affected by the generation of fermentation inhibitors which are process specific. Also there might be an effect of the potential presence of residual solvent (Ionic liquid) in the hydrolysates. This work primarily focuses on the effect that the ionic liquid pretreatment step will have on the fermentation of biomass with different ethanologenic yeasts like Saccharomyces Cerevisiae and, Pichia stipitis. Also optimization of the fermentation process with regards to the Separate hydrolysis and Fermentation (SHF) as well as Simultaneous Saccharification and Fermentation (SSF) will be studied.

1.Anantharam P Dadi, Sasidhar. Varanasi, and Constance .A. Schall,, Enhancement of cellulose saccharification kinetics using an ionic liquid pretreatment step. Biotechnol. and Bioengg, 95 (5) 2006 904-910. 2.Anantharam P Dadi, Sasidhar. Varanasi, and Constance .A. Schall,, Mitigation of cellulose recalcitrance to Enzymatic hydrolysis by ionic liquid pretreatment. Applied Biotechnol. and Bioengg, 136-140 2007 407-421.

513l Reaction Kinetics of Hydrothermal Treatment of Lignin

Bo Zhang, BioTechnology Institute, University of Minnesota, 1479 Gortner Avenue, Suite 140, St. Paul, MN 55108 and Hua-Jiang Huang, Department of Bioproducts and Biosystems Engineering, University of Minnesota, 2004 Folwell Ave, St. Paul, MN 55108.

Lignins derived from abundant and renewable resources are nontoxic and extremely versatile in performance, qualities that have made them increasingly important in many industrial applications. We have shown recently that liquefaction of lignin extracted from aspen wood resulted in a 78% yield of bio-crude oil. In this paper, the direct conversion of five types of lignin was studied, including Kraft pine lignin provided by MeadWestvaco Corporation, Kraft pine lignin obtained from Sigma-Aldrich, organosolv lignin extracted from oat hull, the residues of mixed southern hardwoods and switchgrass after hydrolysis. The major compounds from lignin liquefaction were identified by GC-MS. A kinetic model was proposed to describe hydrothermal treatment of Kraft pine lignin, and compared with another model from literature. The kinetic parameters of the presented model were estimated, including the reaction constants, the pre-exponential factor and the activation energy of the Arrhenius equations. Results showed that the presented model is well in agreement with the experiments.

513m Planning A Sustainable Supply Network for Feedstock to A Typical Biorefinery

Young Kim[1], Choamun Yun[1], Sunwon Park[1], and L. T. Fan[2]. (1) Department of Chemical and Biomolecular Engineering, Korea Advanced Institute of Science and Technology, Daejeon, South Korea, (2) Department of Chemical Engineering, Kansas State University, 1005 Durland Hall, Manhattan, KS 66506

Biorefineries have been increasingly gaining interest as an alternative to oil refineries. The former produces fuels and chemicals from biomass, which is renewable, and the latter from petroleum, which is not renewable. The consumption of fuels and chemicals from biomass appears to be the only legitimately viable means of reducing CO2 release to the atmosphere, thereby minimizing global warming. Unlike the majority of oil refineries, the raw materials, i.e., feedstock, for a typical biorefinery tends to come from wide-ranging sources, which must be pretreated differently. This poses a unique set of challenges in configuring the sustainable supply-chain network, or simply supply network, for the feedstock to the biorefinery. To meet such challenges, a highly robust and efficient method has been developed to design or synthesize the optimal and alternative supply networks for the biorefinery. The method's efficacy is demonstrated by applying it to a realistic example.

513n Biodiesel from Coconut Oil- an Experience in Tonga

Randy S. Lewis, Allyson Frankman, Jacob Jones, and W. Vincent Wilding. Department of Chemical Engineering, Brigham Young University, 350 Clyde Building, Provo, UT 84602

With engineering increasingly becoming an international discipline, engineering programs are enhancing opportunities to prepare students for such a global industry. Being an engineer of tomorrow's world will require students to understand and work with different cultures, peoples, practices, ethics and paradigms. In addition to successful careers, opportunities for engineering graduates to apply their expertise to solve both technical and social problems in the world around them are beneficial.

In order to engage engineering students in global opportunities, a 3-credit hour course was recently developed that involved chemical, mechanical, civil/environmental, and electrical/computer engineering students. The first course was held during the Winter 2007 semester and was followed by a trip to Tonga in May 2007 with 26 students. Students ranged from sophomores to seniors. The trip involved the implementation of a biodiesel production process using coconuts. There is an abundance of coconuts in Tonga, providing great potential to offset the high import costs of diesel. Most of the energy used in Tonga originates from diesel fuel processes.

The biodiesel process involved the entire scope from grinding coconuts, extracting the oil, converting the oil to biodiesel, washing and utilizing the biodiesel, recovering methanol, making soap from the glycerin, and appropriately discharging the waste water. The batch process utilized a 40 gallon reactor. The process was demonstrated to high school students and government officials. During the course, students worked on various aspects of improving the process. An economic analysis and business plan were also completed.

The entire process will be discussed. Included in the discussion are technical aspects and challenges of implementation in Tonga, technical areas that can be improved, safety and environmental aspects, societal aspects, future opportunities, and economic

REDUCTION IN TAR YIELD BY TEMPERATURE PROFILES OF GAS AND SOLID PHASES DURING PYROLYSIS OF WASTES UNDER COUNTERCURRENT FLOW

Yoshikage Ohmukai, Isao Hasegawa, Kazuhiro Mae
Kyoto University, Kyoto-daigaku katsura, Nishikyo-ku, Kyoto, 615-8510, Japan

Introduction

Pyrolysis process is one of the effective conversions for biomass. Many studies have been made to suppress tar formation from biomass during pyrolysis [1−3]. We have proposed to use a shaft kiln in which the flow of gas and solid phase is countercurrent and we investigated pyrolysis and steam reforming reaction of volatile from biomass and several polymers with various temperature profiles in the kiln. High temperature gas is fed to the kiln from the bottom and low temperature solid waste is fed from the top. The solid phase is heated by the high temperature gas phase without external heating from the furnace wall so that the temperatures at which the volatile is evolved from solid waste in pyrolysis and secondary gas phase reactions proceed are different. The features of countercurrent flow affect product distribution through gas phase reactions in the kiln. Moreover, the features are dependent on operating conditions of the kiln and can be controlled separately at quite different profiles, which enables us to design a suitable temperature profile of gas phase for decomposing the tar evolved from solid phase. For utilizing the features of the shaft kiln and developing an energy recovery process from municipal waste, we pyrolyzed the model compounds using a two-stage reactor controlled at different temperatures so as to identify the reaction behavior in terms of the temperature profile in the shaft kiln. We examined the product distribution of the exit of the reactor under various combinations of temperature profiles in each stage of the reactor. From the obtained product distributions, kinetic analysis for the secondary gas phase reaction was performed, and a lumped kinetic model describing gas phase reaction in the reactor was developed.

Experimental section

Samples

The biomass used in this study is Japanese apricot, and the plastics are low density polyethylene (PE), polypropylene (PP), and polystyrene (PS).

Pyrolysis of solid phase and secondary gas phase reactions

Sample was pyrolyzed using the two-stage tube reactor shown in Fig. 1 without any catalyst. The reactor tubes were made of SUS 316 tube of 10.7 mm i.d., and both of the heating regions were 140 mm in length. Atmospheric N_2 gas was fed into the reactors from the bottom of the lower reactor at the rate of 0.33−0.53 ml (STP) s^{-1}. The empty upper reactor was heated at 700, 800 or 900°C prior to pyrolysis. The lower reactor that contained ca. 0.3 g sample was heated at the rate of 30°C min^{-1} up to 600°C and kept at the temperature for 3 min to pyrolyze the sample. Evolved gases and tar passed through the upper reactor and they were pyrolyzed secondarily at higher temperature, and char remained in the lower. Tar and water were collected in an ice trap filled with glass wool. Hydrocarbon gases (HCG), carbon oxides (COx), water, and hydrogen passing though the trap were collected in a gas bag, and their yields were determined using gas chromatograph (Shimadzu, GC-14A) equipped with Porapak TypeQ column or Molecular Sieve 5A. The yields of char and coke were determined using gravimetric analyses of the lower and upper reactor, respectively. Water was determined based on the gas analysis using GC-TCD and tar analysis using Karl-Fischer moisture titrator (Kyoto Electronics, MKS-510N). Tar was determined by difference. Some of the tar composition was analyzed using gas chromatograph equipped with OV-17.

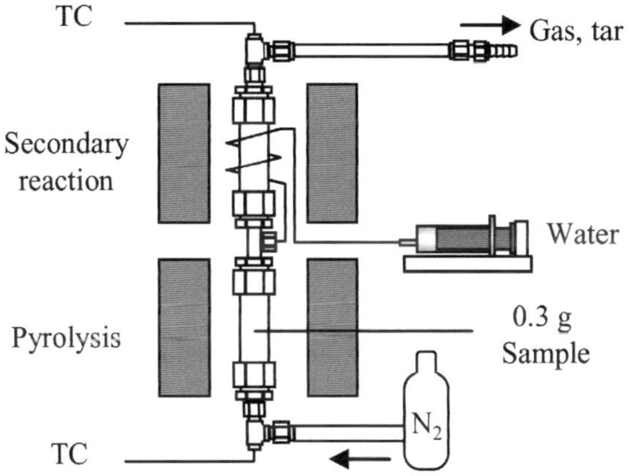

Fig.1 Schematic diagram of two-stage tube reactor. Water feed pump is removed and joint is plugged when pyrolysis is performed.

Results and discussion

Pyrolysis of plastics

Fig. 2 illustrates product distribution in slow heating pyrolysis of plastics on each temperature. Most of sample was decomposed into tar, and a small amount of hydrocarbon gases and coke were obtained through flash pyrolysis of each plastic using CPP at 590˚C. However, tar was decomposed, and formation of hydrocarbon gases was promoted when plastics were pyrolyzed using two-stage reactor because secondary gas phase reactions proceeded in it. Yellow liquid tar was obtained and a large amount of coke was formed as a deposit on the internal surface of the upper reactor through pyrolysis of volatile from PE or PP at higher temperature. Meanwhile, the tar of PS was burnt umber and accompanied by fine-grain coke to the ice trap. The difference implies that each coke was formed through different reaction mechanisms. That is, coke was derived from hydrocarbon gases, especially ethylene in the case of PE and PP while it was from a condensation reaction of benzene ring in tar in the case of PS. These mechanisms are also explained from the product distribution; gas yields were lower and coke yield was higher in pyrolysis of volatile from PE and PP at 900˚C than that at lower temperature while coke was formed without decrease of gaseous products for PS.

Pyrolysis of mixed plastics was performed to examine the effect of mixing on product distribution. Generally, it is known that decomposition reaction is promoted during pyrolysis at ca. 500°C by adding PS to other plastics [4]. However, no significant effect of mixing on hydrocarbon gases yield was observed except for ethylene through pyrolysis at 700–900°C. When mixture of PS and other plastics was pyrolyzed at 900°C, coke and hydrogen yields were lower and ethylene yield was higher than the case where each plastic was pyrolyzed separately at the temperature. The mixing effects are explained from coke formation reaction given by reaction (VIII) in Table 2. The coke formation follows the second order reaction kinetics while the other pyrolysis reactions are of the first order. The volatile including ethylene derived from PE is diluted with that from PS, and the coke formation is suppressed by adding PS to PE because PS volatile contains little ethylene.

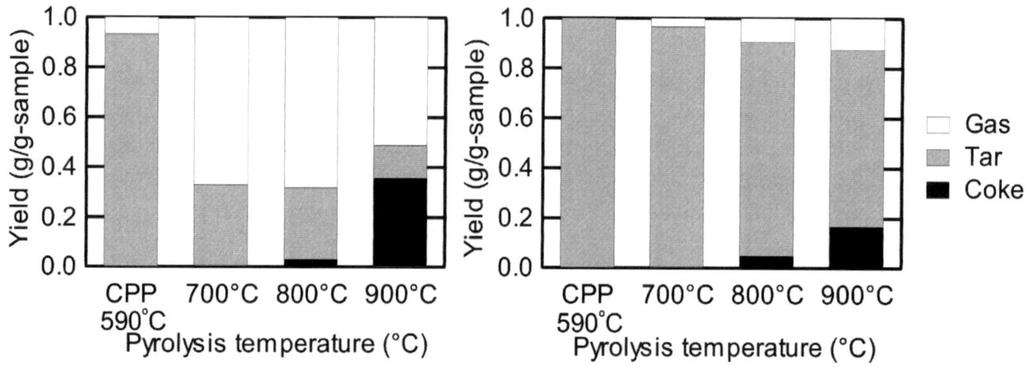

Fig.2 Product distribution on pyrolysis of plastics; (a) PE, (b) PS.

Pyrolysis of biomass

Pyrolysis and steam reforming of volatile from Japanese apricot was performed using the two-stage reactor as well as that of plastic. Although we investigated the effect of steam feed rate on distribution of volatile obtained by steam reforming of it at 800°C, the ratio between each gaseous product including H_2 and CO was constant. This result implies that the steam reforming does not proceed at 800°C or sufficient steam is evolved during the pyrolysis of biomass to decompose pyrolytic tar from itself through the steam reforming. Judging from above, the effect of steam feed rate on gaseous product distribution through steam reforming of biomass tar is negligible. Consequently, pyrolysis of biomass at 700–900°C was carried out without steam feeding as shown in Fig. 3. Char represents the residue remained in the lower reactor and volatile is partially converted to coke in the upper. The pyrolysis of volatile from biomass at high temperature increased CO yield and decreased water and tar yield. These results indicate that steam reforming of biomass tar proceeded by consuming water derived from its own pyrolysis without any catalyst. Moreover, since the yield of hydrocarbon gases increased with temperature, not only steam reforming of tar but also further pyrolysis proceeded at this temperature range.

To analyze pyrolysis behavior of biomass in detail, the flash pyrolysis of volatile from biomass was performed. Fig. 4 shows gaseous products distribution obtained through flash pyrolysis of it at 500–700°C. CO and C_5 gas yield significantly increased at 700°C while there was almost no change in any component up to 600°C. CO would be produced by decomposition of functional groups with oxygen and C_5 gas would be produced by cleavage of ring structure of cellulose monomer because it has a large monomer structure. Considering monomer molecule of cellulose consists of six carbons, however, it is strange that C_5 molecule was produced rather than C_6 one. We considered the reason for this phenomenon in terms of not only the cleavage of ring structure of the tar but also decomposition of it proceeded during the pyrolysis of tar as follows. The ring structure of D-glucose is cleaved in hydrolysis, and subsequently CH_2OH group of either C_1 or C_6 atom is decomposed in pyrolysis, or cleavage of ring structure in pyrolysis leads to formation of CO by consuming C_1 and O atom that C_1–O–C_5 bond is composed of through radical reaction. H_2O is so reactive at 700°C that steam reforming of tar proceeds by reacting with radical molecule produced through pyrolysis of tar at the temperature while it does not proceed in case water coexists with tar at ca. 350°C during conventional pyrolysis. For the reasons just mentioned, thermal instability of the tar at high temperature based on its molecular structure causes decomposition in steam reforming at 700°C without any catalyst, unlike usual steam reforming of toluene. This result showed the feasibility that further decomposition and removal of the tar will be achieved through setting temperature distribution of the reactor at high temperature so as to promote secondary pyrolysis of it unless employing catalysis in tar reforming.

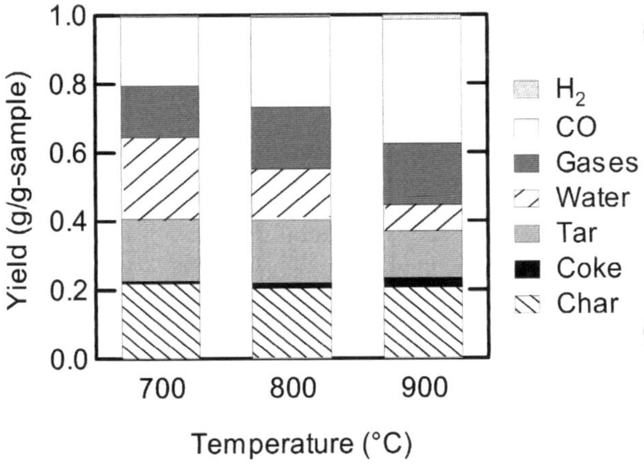

Fig.3 Effect of the pyrolysis temperature on product distribution for pyrolysis of Japanese apricot using two-stage reactor

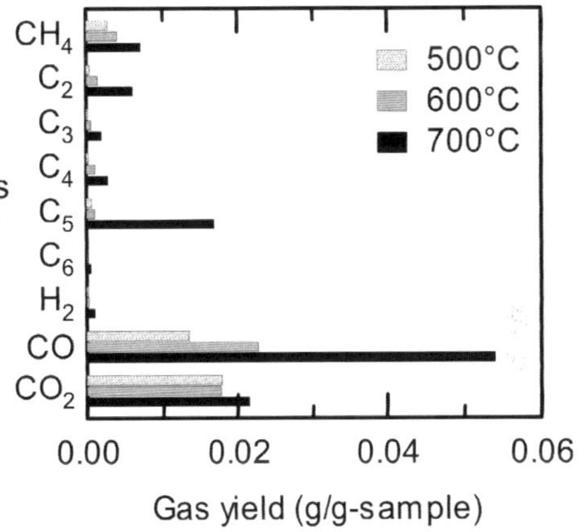

Fig.4 Gaseous product distributions in flash pyrolysis of Japanese apricot at 500 to 700°C using two-stage reactor

Pyrolysis model of mixture of biomass and plastic

Reaction stoichiometry

First, we explain the way to develop reaction stoichiometry in a gas phase reaction model. Since it is practically impossible to develop a kinetic model that takes into account all the reactions that occur simultaneously during the process, pyrolysis is often studied in terms of lumped kinetic models. Many of biomass models are based on the assumption that the interactions between the components are not significant. The results shown above also indicate that there is no significant synergistic effect of plastics during pyrolysis under the condition that pyrolysis was performed in this study, and the mutual effect of plastic and biomass is not apparent, either. Considering these tendencies, the global pyrolysis behavior is described by the algebraic addition of individual pyrolysis behavior, corresponding to the decomposition of PE, PP, PS, and biomass. Based on the experimental results obtained in this study, a pyrolysis model was developed by taking into account 12 components: aliphatic tar from polyolefin, aromatic tar from PS, biomass tar, BTX, CH_4, C_2H_4, hHCG (heavier hydrocarbon gases, ethane and C_3-C_5), H_2, H_2O, CO, CO_2, and coke. For simplicity, the aliphatic tar is represented by $C_{12}H_{24}$, the aromatic tar is by C_8H_8, the biomass tar is by $C_6H_4O_2$, BTX is by C_6H_6, and hHCG including ethane is by C_3H_6, respectively. The composition of the biomass tar was speculated from the mass balances of each element of the products during pyrolysis.

Since both pyrolyses of biomass and plastic is described as two steps of primary pyrolysis and secondary gas phase reaction, the reaction paths between the components were proposed as shown in Fig. 5, where the broken lines are for the primary reaction and the solid lines are for the secondary gas phase reaction. The reaction stoichiometries for the reaction paths were assumed to be formulated as shown reactions (1) and (I)–(XIII) in Table 1 where S_m denotes the dry solid, P_j the products, m = PE, PP, PS and biomass, and j = $C_{12}H_{24}$, C_8H_8, $C_6H_4O_2$, C_6H_6, CH_4, C_2H_4, C_3H_6, H_2, H_2O, CO, CO_2 and char (char is only for biomass). Meng et al [5] presented a lumped model to predict yields of pyrolysate on pyrolysis of heavy oils although it was aimed at catalytic pyrolysis. We lumped light hydrocarbon gases and H_2 in different way from their study because we performed pyrolysis at higher temperature, and the ratio of heavier hydrocarbon gases to H_2 and CH_4 is lower than that of their study. Therefore, it is favorable for us to adopt a model that focuses more on lighter components so as to describe product distribution accurately

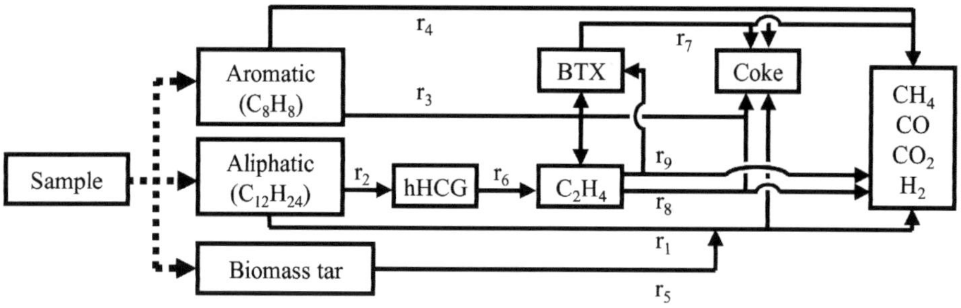

Fig.5 Reaction paths on pyrolysis of mixture in proposed lumped kinetic model

Determination of Kinetic Parameters

The values of the frequency factors and activation energies in the lumped kinetic model are determined. The values of E_{13} and E'_{13} were obtained from the literature [6]. The other rate parameters were estimated so as to give the best fit between the experimental data and the calculated curves using above design equations. Table 2 shows determined values of k_{0i} and E_i. Fig. 6 shows the effect of the temperature on the product distribution from PE and biomass as an example. The experimental data and the calculated curves are in close agreement for most products. However, the fits of some of low-yield components, especially CO_2 from biomass and C_6H_6 from PE, may not be so good. This is because high-yield products were made more account of than low-yield products, which is a necessary

Table.1 Reaction paths, f_{jm} values of tar and water, rate equations, and design equations

Primary reactions

$S_m \rightarrow P_j$ (1)

Secondary reactions

$C_{12}H_{24} \rightarrow 6C + 6CH_4$ (I)

$C_{12}H_{24} \rightarrow 4C_3H_6$ (II)

$C_8H_8 \rightarrow C_6H_6 + 1/2C_2H_4 + C$ (III)

$C_8H_8 \rightarrow 6C + 2CH_4$ (IV)

$C_6H_4O_2 \rightarrow 2CO + CH_4 + 3C$ (V)

$C_3H_6 \rightarrow 3/2C_2H_4$ (VI)

$C_6H_6 \rightarrow 6C + 3H_2$ (VII)

$3C_2H_4 \rightarrow 6C + 6H_2$ (VIII)

$3C_2H_4 \rightarrow C_6H_6 + 3H_2$ (IX)

$C_{12}H_{24} + 12H_2O \rightarrow 12CO + 24H_2$ (X)

$C_8H_8 + 8H_2O \rightarrow 8CO + 12H_2$ (XI)

$C_6H_4O_2 + 4H_2O \rightarrow 6CO + 6H_2$ (XII)

$CO + H_2O \rightleftarrows CO_2 + H_2$ (XIII)

Equations for f_{jm} values

$$f_{C_{12}H_{24}PE,PP} = \left[1 - \left(16 f_{CH_4} + 28 f_{C_2H_4} + 42 f_{C_3H_6}\right)\right]/168 \qquad (2)$$

$$f_{C_8H_8PS} = \left[1 - \left(16 f_{CH_4} + 28 f_{C_2H_4} + 42 f_{C_3H_6} + 78 f_{C_6H_6}\right)\right]/104 \qquad (3)$$

$$f_{C_6H_4O_2biomass} = \left[n_C - \left(f_{CO} + f_{CO_2} + f_{CH_4} + 2 f_{C_2H_4} + 3 f_{C_3H_6} + 6 f_{C_6H_6} + f_{char}\right)\right]/6 \qquad (4)$$

$$f_{H_2Obiomass} = \left[n_O - \left(f_{CO} + 2 f_{CO_2} + 2 f_{C_6H_4O_2}\right)\right] \qquad (5)$$

Rate equations

$r_i = k_i C_s$ (6)

$r_8 = k_8 C_{C_2H_4}{}^2$ (7)

$r_9 = k_9 C_{C_2H_4}{}^2$ (8)

$r_{10} = k_{10} C_{C_{12}H_{24}} C_{H_2O}$ (9)

$r_{11} = k_{11} C_{C_8H_8} C_{H_2O}$ (10)

$r_{12} = k_{12} C_{C_6H_4O_2} C_{H_2O}$ (11)

$r_{13} = k_{13} C_{CO} C_{H_2O}$ (12)

$r'_{13} = k'_{13} C_{CO_2} C_{H_2}$ (13)

The rate constant

$k_i = k_{0i}\exp(-E_i/RT)$ (14)

Design equations

$$\frac{dF_j}{dz} = \sum_{i=1}^{13} a_{ij} r_i S \qquad (15)$$

$$z = 0;\ F_j = F_{0j} + \sum_m f_{jm} W_m(t) \qquad (16)$$

approximation to predict product distribution practically. The curious decrease at 900°C in the C_2H_4 yield from PE is well represented by the model. This decrease and the increases at the temperature in H_2 and coke yield from PE are due to the coke formation from ethylene. Moreover, an almost linear decrease in hHCG yield from PE is well reproduced. Gas yields such as H_2, CH_4, and CO from biomass except for CO_2 are also well simulated by the model. Because the magnitude relation between the activation energies of each reaction in the model is adequate, there is no important error in the proposed network. Thus, the presented simulation model was validated.

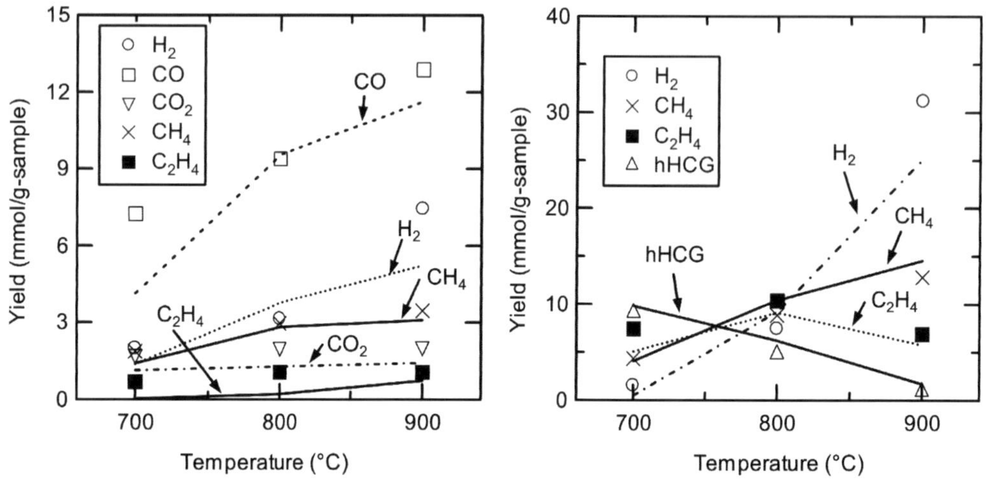

Fig.6 Experimental versus model results for two-stage reactor at different heating temperature: (a) Japanese apricot; (b) PE

Table2 Values of kinetics parameters for a lumped kinetic model obtained using fitting

k_1	1.19×10^9	182
k_2	5.05×10^4	91
k_3	2.55×10^9	223
k_4	9.08×10^3	106
k_5	2.54×10^6	132
k_6	3.34×10^5	111
k_7	3.03×10^4	96
k_8	1.08×10^8	172
k_9	1.62×10^5	132
k_{10}	2.38×10^{10}	226
k_{11}	6.72×10^{10}	230
k_{12}	2.12×10^7	172
k_{13}	6.17×10^2	115
k'_{13}	2.15×10^4	147

Conclusion

In this study, we have proposed to apply shaft kiln of countercurrent flow to pyrolysis of municipal waste and investigated reaction behavior of biomass and plastics in the temperature profile using a two-stage reactor. Main feature of the shaft kiln is that temperatures of gas and solid phase are significantly different from each other, which enables gas phase reactions to be separated from solid phase pyrolysis and shows a feasibility to control reactions in both phases and product distribution actively.

Pyrolysis of volatile from PE, PP, and PS was investigated to analyze reaction behavior and to determine the synergistic effects between plastics. There was no interaction between any plastics to promote decomposition of other at 700–900°C. Tar derived from pyrolysis of plastics was decomposed during steam reforming at 900°C without any catalyst, which implies the feasibility that steam reforming of tar will proceed and tar will be decomposed in real shaft furnace consuming water produced during pyrolysis of biomass or externally provided. The ring structure of biomass tar was split during high-temperature flash pyrolysis. This result shows that the decomposition causes formation of radical molecule and steam reforming of biomass tar would proceed consuming water produced during pyrolysis of itself.

A lumped kinetic model describing gas phase reaction during pyrolysis and steam reforming of volatile from biomass and plastics has been developed based on the experimental results. The simulation model based on the reaction path network from experimental results was found to be successful in reproducing the product distribution. Then, the model was expected to be utilized for predicting the product distribution.

References

1. Lappas AA, Samolada MC, Iatridis DK, Voutetakis SS, Vasalos IA. "Biomass pyrolysis in a circulating fluid bed reactor for the production of fuels and chemicals". *Fuel* 2002; **81**:2087-95
2. Samolada MC, Baldauf W, Vasalos IA. "Production of a bio-gasoline by upgrading biomass flash pyrolysis liquids via hydrogen processing and catalytic cracking". *Fuel* 1998; **77**:1667-75
3. Triantafyllidis KS, Iliopoulou EF, Antonakou EV, Lappas AA, Wang H, Pinnavaia TJ. "Hydrothermally stable mesoporous aluminosilicates (MSU-S) assembled from zeolite seeds as catalysts for biomass pyrolysis". *Microporous Mesoporous Mater* 2007; **99**:132-9
4. Ciliz NK, Ekinci E, Snape CE. "Pyrolysis of virgin and waste polypropylene and its mixtures with waste polyethylene and polystyrene". *Waste Manage* 2004; **24**:173–81
5. Meng X, Xu C, Gao J, Li L. "Catalytic pyrolysis of heavy oils: 8-lump kinetic model". *Appl. Catal.*, A 2006; **301**:32-8
6. Maruyama H, Kato S, Meguri N, Takeda M, Ueda A, Yamada R et al. "Shift reaction of product gases in entrained flow coal gasifier". In: *SCEJ Proc. Energy Eng. Symp.*, Tsukuba: Division of energy engineering, the society of chemical engineers, Japan; 2001; **1**: p. 3-6

Transport Phenomena with the Heterogeneous Chemical Reactions Caused in the Pyrolysis of Biomass Particles

Tatsuya Hinauchi, Ken-ichiro Tanoue, Tatsuo Nishimura,
Yamaguchi University, Ube, Japan

Miki Taniguchi and Ken-ichi Sasauchi,
Chugai Ro Co.ltd, Sakai, Japan

Introduction

Recent years, it needs quickly to produce the environment-friendly energy resources. Especially, people are interested in the production of woody biomass energy by unused forest [8,9,14,15]. The biomass is renewable, abundant, storable, substitute and carbon – neutral resource. In Japan, it needs to propagate a compact gasifier, which has high quality and corresponds with the amounts of collected biomass and the demand of energy. As Japan has many mountains with steep slopes, it is difficult to transport the felled trees. So that, the development of a compact gasifier of high quality, which can cope with variations both in the amount of biomass collected and in energy demand, is necessary. Furthermore, in order to ensure that the system is used in the most efficient manner, it is necessary to increase our understanding of the reaction mechanisms involved in both decomposition of biomass and heat transfer in the biomass layer.

Pyrolysis of biomass consists of many interdependent reactions; nevertheless, it can be reduced to the reaction illustrated in figure 1, which is universally known as the Broido-Shafizadeh mechanism [2,11]. In the primary reactions, biomass decomposes to gas, tar and char. This reaction has been studied extensively in the form of cellulose decomposition. In the secondary reaction, the tar generated in the primary reaction decomposes to light gas and char. However, as the evaporation process is not included in the mechanism, it is difficult to understand whether the decomposition is the endothermic process or the exothermic one. Pyrolysis rates in woody biomass have been investigated using various methods,

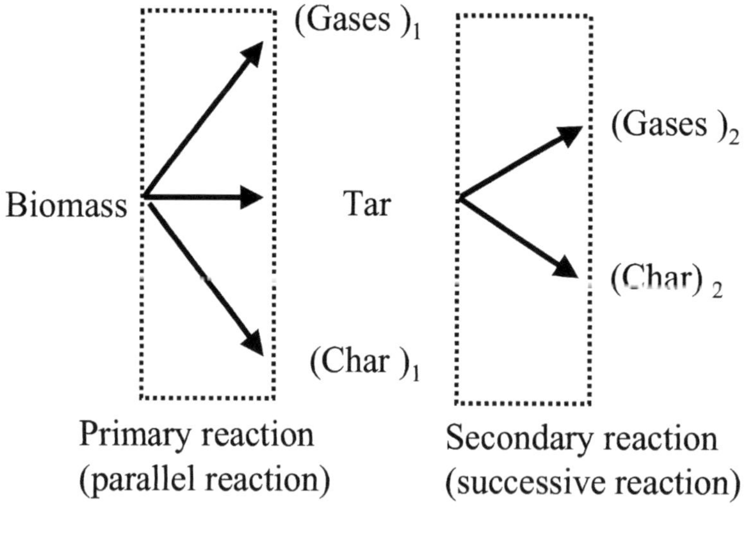

Figure 1. Previous reaction mechanism of pyrolysis of woods[2,11] .

including thermogravimetric analysis [5,7,11,13], analysis of weight change among the main components of the product mixture (gas, tar and char) [3,4,10], and analysis of heat and mass transfer during the reaction [1,6]. In all of these studies, the kinetics of pyrolysis were analyzed using samples with weights of particle diameters of a few tens of microns in diameter. However, in a real pilot plant, sample sizes in grams and biomass particles with an average diameter of several mm have been utilized [8]. In order to estimate the pyrolysis rate based on the results obtained [1, 3-7, 10, 11, 13], an understanding of heat transfer not only in the biomass particles but also in the biomass layer is required.

The purpose of this study is that the model of the kinetics in the pyrolysis at the macroscopic point can be presented by no taking account of the elementary chemical reactions. Concretely, the relationship between the generation of the gas and the temperature of the biomass layer and the material balance of the generated components have been investigated in the rapid pyrolysis of the sawdust of Japanese cender.

Experimental procedure

Figure 2 shows the test rig for pyrolysis experiments of woody biomass. The rig consists of the supplying of nitrogen gas, a tube reactor, a cold trap for tar and water, gas-sampling bag and gas flow meter. The nitrogen gas introduces into the reactor until the concentration of O_2 at the exit of the reactor is less than 1 %. After that, the supplying of the nitrogen gas stops. The tube wall is heated by the electric furnace until the setting temperature, T_S. The temperature of the wall is maintained for about 30 min. The cage with the biomass of W_0 = 0.6 g introduces into the reactor and maintains for 300s. The cage is supported by the K-type thermocouple (Yamari Co., Ltd., ϕ = 1mm). The temperature of the biomass layer is measured by the thermocouple and recorded in a memory of the personal computer through the data logger (Keyence Co., Ltd, NR-1000). The heating rate of the biomass layer is estimated by the following equation.

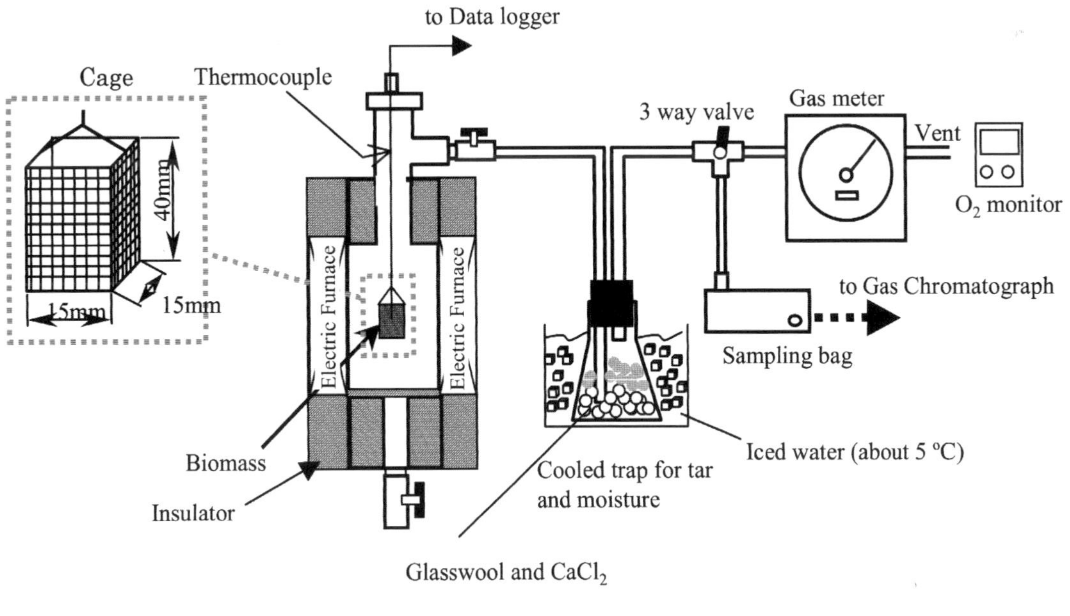

Figure 2. Experimental test rig for the pyrolysis of woody biomass.

$$\dot{T}_b = \frac{T_S - T_0}{\Delta t|_{TS\pm0.5}} \qquad (1)$$

Where T_S, T_0 and $\Delta t|_{TS\pm0.5}$ show the setting temperature, initial temperature of the biomass layer and the elapsed time when the temperature of the biomass layer approaches to the setting temperature in the range of about ±0.5 K, respectively. Tar and water are adsorbed by the cooled traps. Generated gas through the traps fills in the sampling bag (GL Science Co., Ltd) for measurement of gas component. It is difficult to measure the gas flow rate because the pyrolysis of the biomass will complete immediately for high temperature. In this work, the gas meter and the timer are recorded simultaneously by a digital video camera (Sony Co., Ltd, DCR-VX9000). Then, the gas flow rate is calculated. In these experiments, sawdust particles of Japanese cendar are used as the test species of the biomass. Before the pyrolysis, the particles are dried by an oven (Isuzu Co., Ltd, BNS-112S) for 120min at 110 °C. The experimental conditions for rapid pyrolysis are shown in Table 1.

Results and discussion

Relationship between heat and mass transfer in pyrolysis of biomass

Figure 3 shows the experimental results of temperature of biomass layer, T_b, and generated gas flow rate, G, for various setting temperature, T_S. For T_S = 200 °C, it takes about 300s that the temperature of the layer attains to 200 °C. The maximum of generated gas flow rate is about 0.06 l / min. For T_S = 400 °C, the attaining time to 400 °C is about 200s. The generated gas flow rate has two maximum values at 30s and 100s. The temperatures of the layer are about 100 °C (boiling point of water) and 300 °C (decomposed temperature of cellulose), respectively. The higher the setting temperature makes the shorter the attaining time of the layer and has the larger the maximum gas flow rate.

Material balance through the pyrolysis of woody biomass

Figure 4 shows the yields of the char, tar+water and generated gas for various setting temperature. Firstly, the change of the char is mentioned as a function of the setting temperature. For T_S = 200 °C, the weight of the char doesn't change from non-pyrolized biomass. For 250 °C < T_S < 300 °C, the biomass is partly carbonized and the yield of the char decreases slightly. For 350 °C < T_S < 400 °C, the carbonization of the biomass progresses dramatically. Especially, for T_S = 400 °C, the yield of the char is about one-third time of that of non-pyrolized biomass. For 400 °C < T_S < 800 °C, the yield of the char decreases slightly with the setting temperature. Secondary, the change of the tar + water is mentioned. For 200 °C < T_S < 400 °C, the yield of the tar + water increases because both the evaporation of the water (T_S < 300 °C) and the decomposition of the cellulose (300 °C < T_S < 400 °C) seem to occur. Finally, the change of the gas is mentioned. For 200 °C < T_S < 300 °C, the main gases (CH_4, CO, CO_2 and H_2) does not generate. For 350 °C < T_S < 450 °C, the yield of the gas increases dramatically because the volume of the gas increases and the main component of the gas is CO_2. On the other hand, for 450 °C < T_S < 800 °C,

Table 1. Experimental conditions.

	Reactor	
Setting temperature	T_S (°C)	200 - 800
Operation time	t (min)	5
	Biomass	
Average particle size	$D_{p,50}$ (mm)	1.1
Initial weight	W_0 (g)	0.6

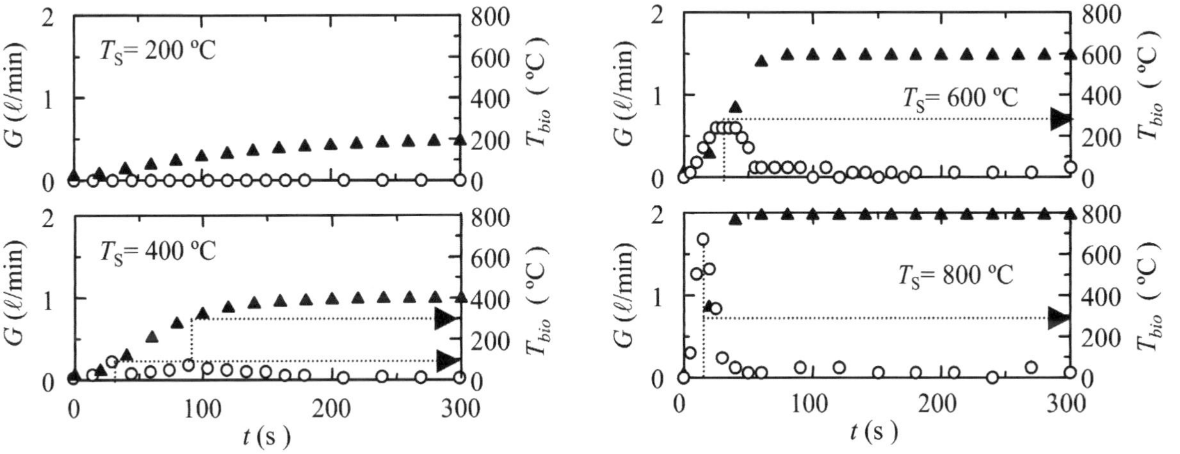

Figure3. Generated gas flow rate and temperature of biomass layer as a function of time.

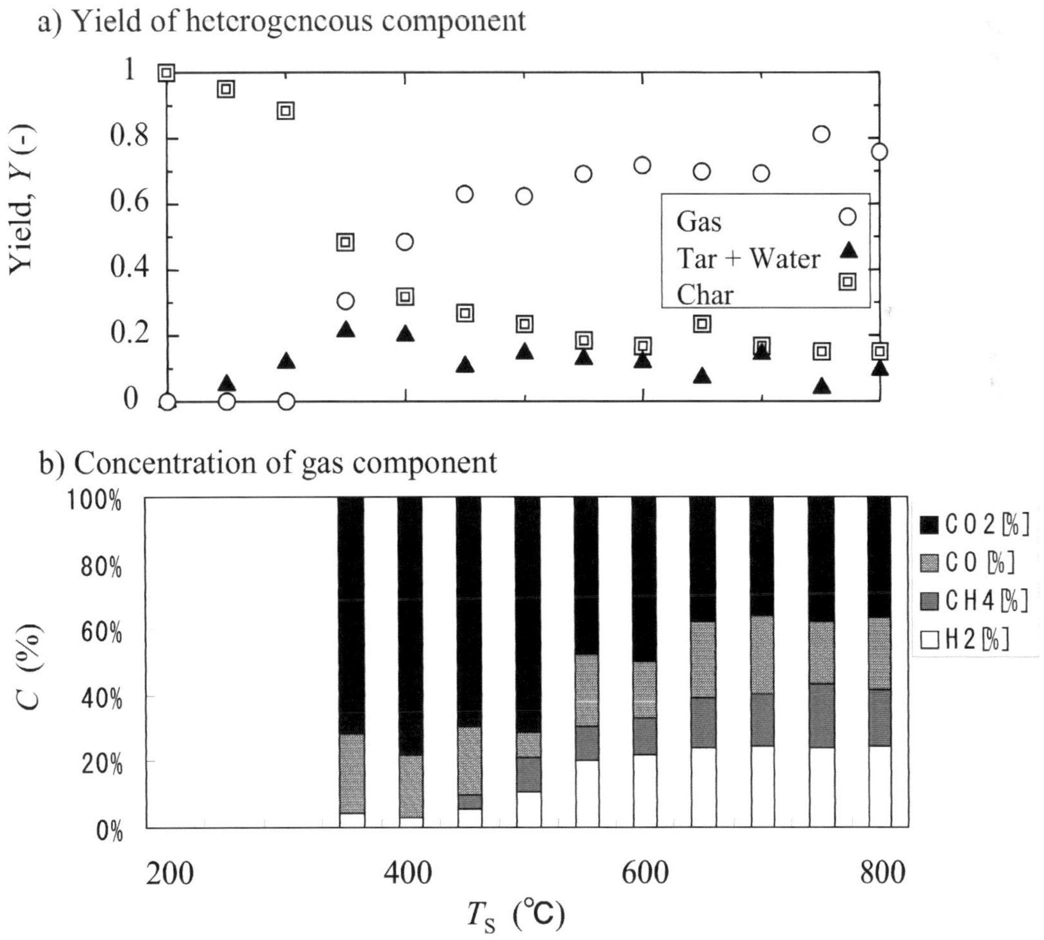

Figure 4. Balance of materials after rapid pyrolysis of woody biomass.

the yield of the gas slightly increases because not only CO_2 but also H_2, CO and CH_4 become generating. Therefore, it seems that the gas with lower molecule mainly generates by the decomposition of tar.

Pyrolysis rate and new chemical reaction model

Figure 5 shows the dependence of the pyrolysis rate on the set temperature. When calculated based on the yield of the tar + water component, the rate was found to increase at temperatures between T_S = 250°C and T_S = 350°C due to decomposition of biomass but decreased at 350°C < T_S < 800°C due to decomposition of the generated tar. Gases other than steam begin to be generated at 350°C. Considering the pyrolysis rate in terms of gas yield, two different rates were observed depending on the temperature, with a dramatic increase at in gas yield at 350°C < T_S < 450°C and a slight increase at 450°C < T_S <

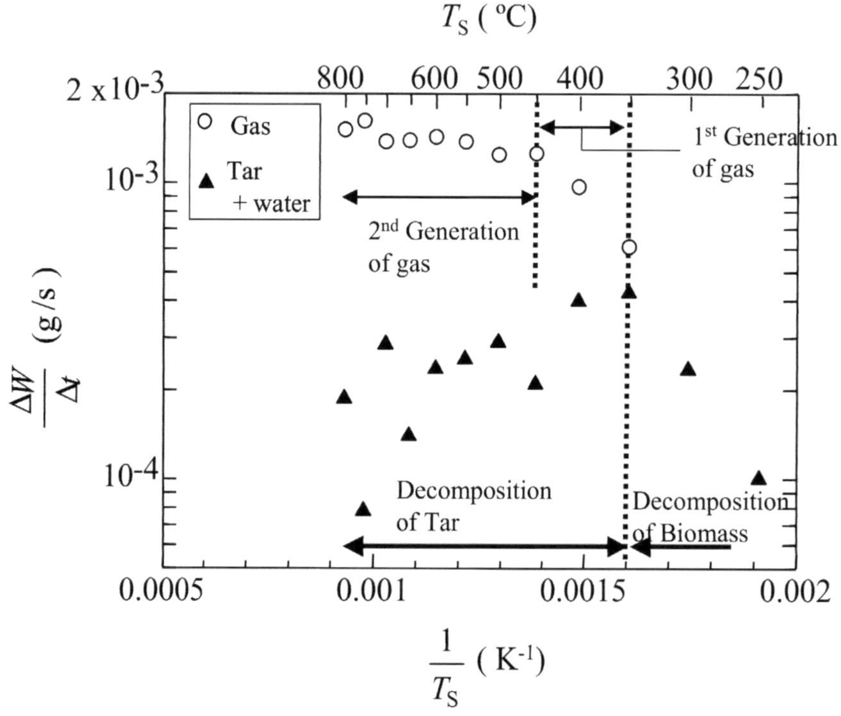

Figure 5. Dependence of rapid pyrolysis rate on the set temperature.

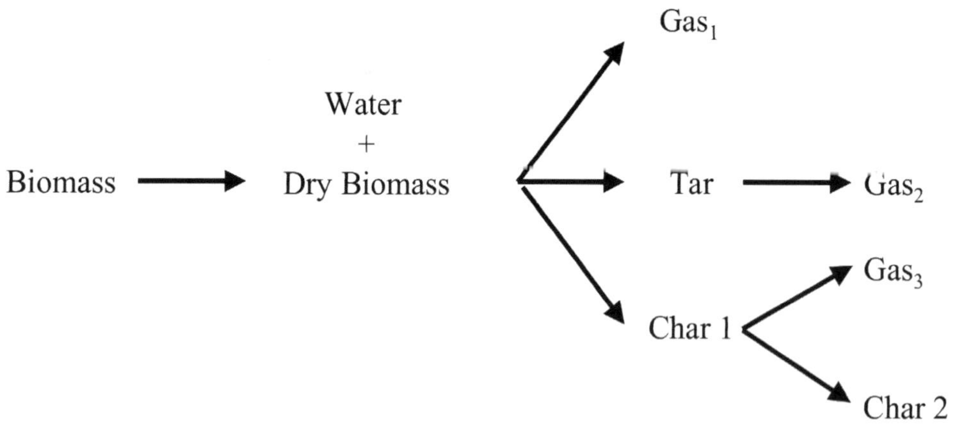

Figure 6. Proposed mechanism for heterogeneous chemical reactions involved in pyrolysis.

800°C. Therefore, it seems that at least two different processes are involved in determining the pyrolysis rate.

From these experimental results, the mechanism of the chemical reactions through the pyrolysis can be explained by Figure 6. Firstly, the evaporation of water from the biomass can occur and then the dried biomass can be generated. Secondly, the dried biomass decomposes to the gaseous products 1, the tar and the lignin. Thirdly, the generated tar decomposes to the gaseous products 2. Finally, the generated lignin decomposes to the gaseous products 3 and the char. In near future, the kinetics of the reactions will be investigated quantitatively.

Conclusions

Transport phenomena in the pyrolysis of woody biomass has been investigated experimentally. Main conclusions are summarized as follows;

1) In the rapid pyrolysis reaction, two different pyrolysis rates were obtained, with different reactions taking place depending on the temperature for a process producing tar + water and a process producing gas. This implies that there are different gas generation processes.
2) The course of the pyrolysis reaction depends strongly on the set temperature, and can be divided into four processes: i) evaporation of water; ii) decomposition of dry biomass; iii) decomposition of generated tar ; iv) decomposition of generated char.

Acknowledgements

This work was supported in part by a Grant-in Aid for Scientific Research C (No. 19560174) from the Japan Society for the Promotion of Science. Our special thanks are extended to Dr. Morihisa Yokota of Ube Industries, Ltd., for helpful discussions on the work.

Refernces

1. Babu, B. V. and Chaurasia, A. S. (2004), "Heat transfer and kinetics in the pyrolysis of shrinking biomass particle," Chem. Eng. Sci., 59, pp.1999-2012
2. Broido, A. (1976), "Kinetics of solid-phase cellulose pyrolysis. In: F. Shafizadeh, K. V. Sarkanenm, D. A. Tillman, editors. Thermal uses and properties of carbohydrates and lignins," pp.19-35, New York, Academic Press
3. Fagbemi, L., Khezami, L. and Capart, R.(2001), "Pyrolysis products from different biomasses: application to the thermal cracking of tar, " *Applied Energy*, 69, pp.293-306
4. Ferdous, D., Dalai, A.K., Bej, S. K., and Thring, R. W.(2002), "Pyrolysis of Lognins: Experimental and Kinetics Studies," *Energy Fuel*, 16, pp.1405-1412
5. Gomez, C. J., Manya, J. J., Velo, E. and Puigjaner, L.(2004), "Further Applications of a Revisited Summative Model for Kinetics of Biomass Pyrolysis," *Ind. Eng. Chem. Res.*, 43, 901-906
6. Koufopanos, C. Papayannakos, A., N., Maschio, G. and Lucchesi, A.(1991), "Modelling of the Pyrolysis of Biomass Particles. Studies on Kinetics, Thermal and Heat Transfer Effects," *Can. J. Chem. Eng.*, 69, pp.907-915
7. Muller-Hagedornm, M., Bockhorn, H., Krebs, L. and Muller, U.(2003), "A comparative kinetic study on the pyrolysis of three different wood species," *J. Anal. Appl. Pyrolysis*, 68-69, pp. 231-249
8. Sasauchi, K.(2005), "Application for Electric Generator by Pyrolysis Gasification of Biomass," *J. Com. Soc. Japan*, 47-139, pp.31-39(in Japanese)
9. Suzuki, T.(2002), "Liquefaction and Gasification of Woody Biomass – Current status of process development," *Mokuzaigakkaishi*, **48-4**, 217-224 (in Japanese)

10. Ueoka, K., Matsushita, Y., Yamamoto, T., Aoki, H., Miura, T.(2004), "Experimental Investigation of Biomass Pyrolysis Mechanism Using Cellulose as a Model Compound," *JASME (B)*, 70-694, pp.1598-1603 (in Japanese)
11. Vamvuka, D., Kakras, E., Kastanaki, E. and Goammelis, P.(2003), "Pyrolysis characteristics and kinetics of biomass residuals mixtures with lignite," *Fuel*, 82, pp.1949-1960
12. Varhegyi, G., Jakab, E., Antal, M. J.(1994), "Is the Broido-Shafizadeh model for cellulose pyrolysis true?, " *Energy and Fuels*, 8, pp.1345-1352
13. Watanabe, T.(2003), "Development of Small-scale Distributed Power Generation System Utilizing Woody Biomass Gasification Technology, " *Taiyoenerugi*, **29-1**, 14-17 (in Japanese)
14. Willner, T. and Brunner, G.(2005), "Pyrolysis Kinetics of Wood and Wood Component," *Chem. Eng. Technol.*, 28-10, pp.1212-1225
15. Yokoyama, S.(2005), "Prospect and Challenge of Woody Biomass in Japan, " *Enerugi to Shigen*, **26-3**, 169-172 (in Japanese)

513q Depolymerization of Lignocellulosic Materials

Gang Huang[1], Hossein Toghiani[2], Charles U. Pittman Jr.[1], and W. Todd French[2]. (1) Chemistry, Mississippi State University, P.O. Box 9595, MS State, MS 39762, (2) Chemical Engineering, Mississippi State University, 330 Swalm Circle President Dr., Mississippi State University, MS 39762

Two types of biomass, switch grass and giant miscanthus, were subjected to pretreatment and acid hydrolysis. The pretreatments used on the grass samples included size-reduction, extraction and concentrated acid impregnation. BET surface area measurements indicated that the size-reduction and extractive-removal by water and ethanol resulted in changed surface areas compared to the as-received biomass samples. Analysis of the as-received and extractive-free samples showed that extractions with water and with ethanol removed inorganic compounds present in the as-received biomass samples. Under the same conditions of acid hydrolysis, the extractive-free samples displayed a greater extent of depolymerization of the cellulose and hemicellulose content compared to the as-received samples. Employing weight ratios of biomass to acid ranging from 1:2 to 1:10, 72% sulfuric acid was used. Examination of experimental results indicated that a biomass/acid weight ratio of 1:3 resulted in the greatest extent of depolymerization for both types of biomass.

554a Comparative Sugar Recovery Yields For Pretreatment Of Corn Stover And Poplar By Leading Technologies Followed By Enzymatic Hydrolysis Of The Remaining Solids

Charles E. Wyman, Chemical and Enviornmental Engineering, University of California at Riverside, College of Engineering Center for Environmental Research and Technology, 1084 Columbia Avenue, Riverside, CA 92507, Bruce Dale, Michigan State University, Department of Chemical Engineering and Materials Science, Room 3247 Engineering Building, East Lansing, MI 48824-1226, Richard T. Elander, NREL, 1617 Cole Blvd, Golden, CO 80401, Mark T. Holtzapple, Texas A&M University, Dept. Chemical Engineering, College Station, TX 77843-3122, Michael Ladisch, LORRE, Purdue University, 500 Central Drive, Potter Building, W. Lafayette, IN 47907, Y. Y. Lee, Chemical Engineering, Auburn University, Auburn, AL 36849, Colin Mitchinson, Genencor International, Palo Alto, CA 94304, and John N. Saddler, The University of British Columbia, 2004 - 2424 Main Mall, The University of British Columbia, Vancouver, BC V6T 1Z4.

Pretreatment is essential to high yields and low costs for biological processing of cellulosic biomass to fuels and chemicals. A team experienced in biomass hydrolysis formed a Biomass Refining Consortium for Applied Fundamentals and Innovation (CAFI) to develop comparative data for leading pretreatment options of ammonia expansion, aqueous ammonia recycle, controlled pH, dilute acid, lime, and sulfur dioxide steam explosion using shared feedstocks, enzymes, procedures, and analytical methods. Controlled sources of corn stover and poplar wood were employed as feedstocks to determine the effect of considerably different classes of cellulosic biomass on performance. Comparative data were developed on sugar recovery for each feedstock during pretreatment and after digestion of the solids produced using different loadings and combinations of enzymes supplied by Genencor International. Standard protocols were applied to close material balances based on this data and used to determine overall sugar release from each combination of pretreatment and enzymatic hydrolysis. All pretreatments recovered high yields of sugars from the hemicellulose and cellulose in corn stover, although high pH technologies tended to be somewhat more effective. However, yields were much more variable for applications of the same technologies to poplar wood, and significant performance differences were observed among these pretreatments for the same poplar variety from two different locations. Overall, these results show the importance of linking selection of pretreatment technology with the feedstock used.

554b Xylanase Supplementation During Saccharification Of Corn Stover And Poplar

Renata Bura[1], Richard Chandra[2], and Jack Saddler[2]. (1) College of Forest Resources, University of Washington, Seattle, WA 98195, (2) Wood Science, University of British Columbia, 2424 Main Mall, Vancouver, BC V6T1Z4, Canada

A commercially-viable biomass to ethanol conversion process should provide maximum sugar recovery (five and six carbons) in a form that can be easily hydrolysed and fermented. In an effort to increase overall sugar recovery, we explored the possibility of using water soluble and insoluble streams of steam exploded corn stover and poplar, supplemented with a mixture of cellulases, beta-glucosidases and xylanases during saccharification experiments. In this study, the steam explosion experiments were performed at 190°C and 5 minutes, and 200°C, 5 min and 3% SO_2 for corn stover and hybrid poplar, respectively. Overall improvements in converting xylan to xylose and cellulose to glucose for water soluble and insoluble streams will be discussed.

554c Lime Pretreatment Of Poplar Wood

Mark T. Holtzapple, Texas A&M University, Dept. Chemical Engineering, College Station, TX 77843-3122 and Rocio Sierra, Chemical Engineering, Texas A&M University, Dept. Chemical Engineering, College Station, TX 77843-3122.

The enzymatic digestibility of poplar wood is enhanced by treating with lime plus oxygen. Two types of treatments have been explored: (1) long-term and (2) short-term. The long-term treatment requires 1 to 2 months at 25 to 65 C with 1-atm air. The short-term treatment requires about 6 h at 150 C with 14-atm oxygen. Because of its high lignin content, poplar wood responds better to the short-term treatment, which is more aggressive. The short-term treatment allows the digestion of 90% of cellulose and 83% of hemicellulose using a moderate enzyme loading of 15 FPU/g glucan in the original biomass.

554d Enzymatic Digestion And Fermentation Of Liquid Hot Water Pretreated Yellow Poplar

Michael Ladisch[1], Youngmi Kim[2], Nathan Mosier[3], Mira Sedlak[3], and Nancy W. Y. Ho[2]. (1) Agricultural & Biological Engineering, Biomedical Engineering & Laboratory of Renewable Resources Engineering (LORRE), Purdue University, Potter Engineering Center, 500 Central Drive, West Lafayette, IN 47907-2022, (2) Laboratory of Renewable Resources Engineering (LORRE), Purdue University, Potter Engineering Center, 500 Central Drive, West Lafayette, IN 47907-2022, (3) Agricultural & Biological Engineering & Laboratory of Renewable Resources Engineering (LORRE), Purdue University, Potter Engineering Center, 500 Central Drive, West Lafayette, IN 47907-2022

Lignocellulosic biomass is recalcitrant to enzymatic hydrolysis due to its structural complexity. Controlled pH, liquid hot water pretreatment of yellow poplar improves its enzymatic digestibility. Liquid hot water pretreatment conditions were optimized for maximum conversion of yellow poplar to fermentable sugars while minimizing sugar loss to degradation products, at an enzyme loading of 15 FPU cellulase/g glucan. At the optimized pretreatment conditions about 44% of acid soluble lignin initially present in the poplar was solubilized into the liquid fraction of the pretreated poplar. Hot washing of the pretreated poplar enhanced glucose yield two-fold as compared to the pretreated but unwashed poplar, by removing the solubilized lignin that can re-attach to the pretreated poplar surface or nonspecifically bind with enzymes, thereby lowering the sugar yields. Enzymatic digestion of the pretreated and hot washed poplar at 15 FPU cellulase per gram of glucan and a solids loading of 150 g dry solid/L and resulted in 50% of glucose and 60% of xylose yields after 48 hours. Fermentation of the hydrolyzate using a xylose co-fermenting yeast strain 424A (LNH-ST) gave a metabolic yield of 0.45g ethanol/g consumed sugars. Possible mechanisms and process conditions for optimizing aqueous hot water pretreatment of poplar.

file:///p|/Print%20on%20Demand%20Titles/00000%20Publishers%20File/AIChE/AIChE%20Annual%202007/All/P95553.HTM1/23/2008 2:54:29 PM

554e Relating Enzymatic Hydrolysis Yields To Substrate Features For Corn Stover And Poplar Solids Produced By Leading Pretreatment Technologies

Rajeev Kumar, Dartmouth College, Hinman Box-8000, Cummings Hall, Thayer School Of Engineering, Hanover, NH 03755, Bruce Dale, Michigan State University, Department of Chemical Engineering and Materials Science, Room 3247 Engineering Building, East Lansing, MI 48824-1226, Richard T. Elander, NREL, 1617 Cole Blvd, Golden, CO 80401, Mark T. Holtzapple, Texas A&M University, Dept. Chemical Engineering, College Station, TX 77843-3122, Michael Ladisch, LORRE, Purdue University, 500 Central Drive, Potter Building, W. Lafayette, IN 47907, Y. Y. Lee, Chemical Engineering, Auburn University, Auburn, AL 36849, Colin Mitchinson, Genencor International, Palo Alto, CA 94304, John N. Saddler, The University of British Columbia, 2004 - 2424 Main Mall, The University of British Columbia, Vancouver, BC V6T 1Z4, and Charles E. Wyman, Chemical and Enviornmental Engineering, University of California at Riverside, College of Engineering Center for Environmental Research and Technology, 1084 Columbia Avenue, Riverside, CA 92507.

Enzymatic digestion data is related to substrate features for solids prepared by leading pretreatment technologies of ammonia fiber expansion, ammonia recycle percolation, controlled pH, dilute acid, lime, and sulfur dioxide pretreatments to gain new insight into mechanisms that account for differences in performance of different pretreatments and feedstocks as part of the cooperative CAFI project. Controlled sources of corn stover, an agricultural residue, and poplar, a woody biomass, are used to gain insight into changes in performance for these two distinct types of cellulosic biomass. Enzymatic hydrolysis is conducted according to standard experimental protocols on the solids from each pretreatment using single lots of commercial and advanced cellulase enzymes provided by Genencor International. The impact of using combinations of different enzyme activities (e.g., hemicellulase) and additives (e.g., non enzymatic proteins, surfactants) on total protein demands is being developed as well. Key features hypothesized to influence solids digestibility are measured and related to hydrolysis performance to account for how different pretreatment technologies, substrates, and enzyme activities interact and gain new insight into key mechanisms that control performance.

554f Logistical Support And Modeling Efforts In Pretreatment Research

Tim Eggeman, Neoterics International, 2319 S. Ellis Ct., Lakewood, CO 80228 and Richard T. Elander, NREL, 1617 Cole Blvd, Golden, CO 80401.

The Biomass Refining Consortium for Applied Fundamentals and Innovation (CAFI) was formed in 1999 as a multi-institutional effort to develop comparative information on cellulosic biomass pretreatment options. Our prior work focused on a comparative evaluation of corn stover feedstock for five leading pretreatment technologies (dilute acid, hot water, ammonia recycle percolation, ammonia fiber explosion and lime). Our current efforts, known as the CAFI II project, have been expanded to cover hybrid poplar as the feedstock and an additional pretreatment (SO2). This paper will describe the logistical support and modeling efforts that the National Renewable Energy Lab has provided to the consortium in the CAFI II project.

574a Lactic Acid Production From Paper Sludge Using Thermophilic Bacteria

Naresh K. Budhavaram and Zhiliang Fan. Biological Systems Engineering, Virginia Polytechnic Institute and State University, 203 Seitz hall, Biological Sysytems Engineering, Blacksburg, VA 24061

Conventionally lactic acid is produced from the starch and sugar sources which not only make the process expensive but also deplete the food reserves. On the other hand lactic acid can be produced at low cost using paper sludge as it is rich in cellulosic content and also does not require any pretreatment. The optimum temperature for the cellulase enzyme is around 50ºC and only thermophiles can sustain at that temperature. The use of thermophilic bacteria in Simultaneous Saccharification and Fermentation (SSF) along with the cellulase enzyme helps in increasing the yield of lactic acid.

Three thermophilic bacterial strains have been screened for their ability to produce lactic acid from the paper sludge by utilizing both cellulose and xylose. Two out of the three have given promising results. The conditions such as temperature and pH will be optimized in the further experiments so as to increase the lactic acid yield. Also the cellulase loading will be optimized. The complex media and the mineral media will be used for growing the cultures and also for fermentations and the results will be compared. Acetic acid and ethanol yields will also be monitored to see the selectivity of the strains in producing lactic acid. Finally batch and semi continuous methods will be compared for the lactic acid yield.

574b Enzymatic Hydrolysis And Fermentation For Afex Treated Corn Stover- An Integrated Biorefinery Approach

Leonardo D. Sousa[1], **Venkatesh Balan**[1], Shishir P. S. Chundawat[1], Derek Marshall[1], Bruce E. Dale[1], and Patricia J. Slininger[2]. (1) Department of Chemical Engineering and Materials Science, Michigan State University, 2527, Engineering Building, East Lansing, MI 48824, (2) Crop BioProtection Research Unit, National Center for Agricultural Utilization Research, USDA Agricultural Research Service, 1815 N. University, Peoria, IL 61604

In the United States, ethanol from corn grain is an established process with an annual production of approximately 4 billion gallons. However, due to increasing demand for renewable fuel and limited availability of corn starch other feedstock (i.e. lignocellulosic agricultural wastes) are being explored. Newer and better pretreatment technologies are emerging to improve digestibility & fermentability of lignocellulosics. One such novel alkaline pretreatment technology is Ammonia Fiber Expansion (AFEX).

In an industrial process it is critical to perform enzymatic hydrolysis at higher solid loadings in order to achieve economical concentrations of ethanol (>4% (w/w)) after fermentation. In this study, corn stover was pretreated using optimal AFEX conditions, followed by enzymatic hydrolysis at varying glucan loadings (1-12%) in a fed batch process. At higher solid loadings the increased concentration of inhibitors (i.e. lignin phenolics, organic acids) hinders enzyme & microbe performance. In order to reduce this inhibitory effect several strategies were explored. (1) Pretreated biomass was washed to remove pretreatment degradation products and (2) Enzymatic hydrolysis was supplemented with protein rich additives. We will report the sugar yield during the process and ethanol yield by performing *Pichia Stipitis* fermentation.

574c Enzyme Engineering of Isolates from Thermoacidophiles for Enhanced Saccharification and Pretreatment Consolidation

Blake A. Simmons[1], Rajat Sapra[2], Diana Roe[2], Jean-Loup Faulon[3], Joanne Volponi[2], and George Buffleben[4]. (1) Sandia National Laboratory, 7011 East Avenue, Livermore, CA 94550, (2) Biosystems, Sandia National Laboratory, 7011 East Avenue, Livermore, CA 94550, (3) Computational Bioscience, Sandia National Laboratory, 7011 East Avenue, Livermore, CA 94550, (4) Energy Systems, Sandia National Laboratory, 7011 East Avenue, Livermore, CA 94550

The goal of this project is to demonstrate that enzyme engineering of cellulases can enable the development of consolidated lignocellulosic ethanol biorefinery unit operations; specifically a consolidation of biomass pretreatment and enzymatic hydrolysis. Our year two efforts are focused on the development, manipulation, and optimization of novel classes of cellulase enzymes that have been isolated from extremophiles. Extremophiles are organisms that exist in unique environments, such as sulfurous calderons and thermal hot springs, and possess enzymes that perform optimally in those environments. The majority of industrial cellulase enzymes to date have been derived from bacteria and fungi. Archaea are the third domain of known life and are finding more applications in biotechnology. This is in no small part because most of the archaeal species identified to date have been identified from extreme environments. Environments such as geothermal and deep-sea volcanic sites (80–121 °C), polar regions (-20 °C), acidic solfatara fields (pH < 4) and alkaline springs (pH > 8) and hypersaline lakes (2–5 M NaCl) which were thought to be inhospitable for life to exist have been shown to be teeming with life forms that are unique sources of enzymes with biotechnological importance.

Evolutionary processes have optimized the structure of enzymes to deliver a desired catalytic function. Proteins, however, are fragile and easily succumb to extreme environmental conditions that are typically associated with established pretreatment technologies for biomass processing, such as dilute acid. We are attempting to modify the sequence, structure, and optimal operating characteristics of these enzymes so that a consolidated bioprocessing approach, collapsing the pretreatment and enzymatic hydrolysis steps into one unit operation, can be implemented to lower costs dramatically. The current estimates of the state-of-the-art pretreatment and enzymatic hydrolysis steps are 25-32% of the estimated total operating costs of a lignocellulosic biorefinery. Three-dimensional structures for each member of the enzyme family are either directly acquired from the protein data base (PDB) or if structural information is not available, homology modeling techniques will be used to build three-dimensional models (if a template structure exists). The active sites of the enzymes are then aligned and compared, and a geometric and chemical "fingerprint" is generated for the site. The active site fingerprint is compared with a set of synthetically accessible scaffolds and/or fragments to identify matches with significant geometric and chemical similarity. The set of scaffolds thus identified are tested and optimized using an iterative process involving detailed atomistic simulations and in-house techniques to suggest modifications followed by the production of new engineered and tailored enzymes for the hydrolysis of cellulosic and lignocellulosic materials.

The hyperthermophilic archaea that thrive under extremes of temperature, pH, salinity etc., have received considerable attention because of the high thermostability of their associated enzymes. The crenarchaeote Sulfolobus solfataricus is a thermoacidophile that grows optimally at approximately 80 °C and pH 2– 4. As such, the proteins isolated from this organism are functional at high temperature and low pH, making them ideal candidates for use in a consolidated dilute acid biomass pretreatment scenario. The genome of S. solfataricus has been sequenced, and three genes (designated sso1354, sso1949 and sso2534) encoding potentially secreted endo-â-glucanases of glycoside hydrolases (GH) family 12 are found in the genome. We selected the SSO1949 gene that generates an endo-â-glucanase (EC 3.2.1.4), exhibits an exceptional activity at extremely low pH, and is thermostable. Our objective is to engineer this enzyme to be a highly efficient catalyst as defined by high turnover and low affinity for binding of the cellulose substrate. This enzyme has a molecular weight of 37 kDa and catalyzes the hydrolysis of carboxymethylcellulose and other oligomers of cellulose to cellobiose and cellotriose. These products can be easily screened for using established commercial assays. We obtained a clone of this gene from the lab of Dr. G. Lipps at the Institute of Biochemistry, University of Bayreuth, Bayreuth, Germany and our current experimental objective is to establish conditions for expression of the proteins, high throughput purification and enzymatic analysis of the protein.

The SSO1949 was cloned into pET-28c-1949Nhis, where the SSO1949 protein carries a fusion of six histidine residues at its N-terminus to enable affinity chromatographic separation of the SSO1949. This plasmid was then transformed into E. coli for expression and protein expression induced via IPTG addition. The active and soluble portion of the purified enzyme demonstrates that the SSO1949 is significantly more active than other commercial endo-â-glucanase at optimal temperature and pH. We will present the results obtained from the computationally predicted sequence changes in terms of improved performance and efficiency.

574d Production of Xylitol from Agricultural Hemicellulosic Biomass

David Demirjian[1], Ryan D. Woodyer[1], Mike Racine[1], and Badal C. Saha[2]. (1) zuChem, 2201 W Campbell Park Dr, Chicago, IL 60612, (2) Fermentation Biotechnology Research Unit, National Center for Agricultural Utilization Research, USDA-ARS, 1815 N University Street, Peoria, IL 61604

The production of value-added co-products from agricultural biomass is an important economic driver for the success of a biorefinery approach to the production of ethanol and other fuels. During most ethanol production methods significant amounts of hemicellulose by-products are produced which are rich in pentoses such as xylose and arabinose as well as other sugars. Xylitol, a five-carbon polyol that has gained acceptance as a natural food sweetener, can be made from pure xylose by chemical hydrogenation, but production from a mixed pentose stream such as that found in hemicellulose is difficult and expensive due to separation issues. We have developed high efficiency fermentation strains capable of converting xylose and arabinose to xylitol by combining a variety of microbial genes and pathways. The fermentation is carried out in an engineered E. coli which also utilizes the other sugars in the mixture as carbon sources, thus producing xylitol with no other competing polyol by-products produced. The increasing demand for xylitol volume should provide a unique match as a co-product made alongside ethanol and other fuels produced from a biorefinery.

Web Page: www.zuchem.com

574e Effect of Continuous Enzymatic Hydrolysis and Simultaneous Saccharification and Fermentation of Pretreated Corn Stover by Bsa Treatment

Michael Studer, Bin Yang, and Charles E. Wyman. CE-CERT Bourns College of Engineering, University of California Riverside, 1084 Columbia Avenue, Riverside, CA 92507

The masking of lignin in pretreated corn stover by addition of non-catalytic proteins (e.g. BSA as the model protein) has been shown to enhance enzymatic cellulose hydrolysis. These additives preferentially attach to lignin and therefore the nonproductive binding of valuable enzymes on lignin is lowered. The optimization of the BSA masking process was studied in batch experiments; parameters of interest were the minimization of the effective BSA concentration and the determination of the necessary contact time to increase hydrolysis rates and yields. These insights were applied to run continuous hydrolysis as well as continuous simultaneous saccharification and fermentation (SSF) experiments. Particular attention is being given to show the effects of BSA and cellulase concentrations, retention time and microorganisms on results. In addition, a mathematical model is being applied to fundamental understanding of how the performance of continuous hydrolysis and SSF and its relationship to key substrate and enzyme features and identify opportunities to enhance high ethanol yields at lower cost.

574f A Novel Approach For Conversion Of Xylose To Ethanol Using Native Yeast Strains

Kripa Rao[1], Sasidhar Varanasi[1], Patricia Relue[2], and Constance Schall[1]. (1) Chemical and Environmental Engineering, University of Toledo, 2801 W Bancroft street, Toledo, OH 43606, (2) BioEngineering, University of Toledo, 2801 W Bancroft street, Toledo, OH 43606

Of the sugars recovered from lignocellulose, wild type Saccharomyces cerevisiae strains can readily ferment most six-carbon sugars. However, xylose, which also forms a major portion of biomass, cannot be fermented by the same native strains of yeast. The pathway of conversion of xylose to ethanol goes via the formation of xylulose, which is a ketose isomer of xylose. There have been many attempts to genetically engineer the yeast strains with over expression of xylose isomerase genes. Our approach however is to convert the xylose to xylulose using the widely manufactured exogenous enzyme Glucose/ Xylose Isomerase (XI) and then ferment the xylulose to ethanol by using different yeast strains like Saccharomyces Cerevisiae D5A and Pichia stipitis in the same reaction vessel. A major obstacle however, to this approach is that the pH at which xylose isomerase displays its optimal activity (pH of 7.5-8.0) is much different from the pH at which the fermentation is carried out. We have successfully demonstrated that the maintenance of two vastly different pH microenvironments in a single vessel was possible. This involves the use of bilayered (XI–urease)immobilized enzyme pellets. The design of the co immobilized pellets and the fermentation parameters leading to maximum ethanol production with different yeasts will be discussed.

603a Performance Of Alkaline Pretreatment As Affected By Reagent, Mode Of Reaction, And Feedstock Composition

Rajesh Gupta and Y. Y. Lee. Chemical Engineering, Auburn University, Auburn, AL 36849

Alkaline pretreatments were applied to hybrid poplar feedstock using aqueous ammonia and NaOH. Two batches of hybrid poplar were used: high lignin (HL) hybrid poplar and low lignin (LL) hybrid poplar. These feedstocks were treated with three different types of alkaline pretreatment processes. Two different modes of operation was used in aqueous ammonia pretreatment: a flow-through process termed as ammonia recycle percolation (ARP), and a batch process named soaking in aqueous ammonia (SAA). Additionally a batch pretreatment using NaOH was investigated. Both reagents were effective in delignifying the biomass and increasing the glucan digestibility. In treatment using ammonia, most of the hemicellulose is retained in solids. NaOH, however, reacts with carbohydrates in biomass during pretreatment causing slight loss in it. Addition of hydrogen peroxide during NaOH treatment was found helpful in increasing the delignification as well as carbohydrate retention in solids. For LL hybrid Poplar, pretreatment by ammonia or NaOH gave above 90% of glucan and xylan digestibilities with use of cellulase/xylanase mixture. For HL hybrid poplar, pretreatment with NaOH gave better results than ammonia. NaOH/H2O2 treated HL hybrid poplar gave nearly quantitative glucan and xylan digestibilities when cellulase/xylanase mixture was applied. The maximum glucan/xylan digestibilities obtained for ammonia treated HL hybrid poplar were 86%/77%. Reaction products in the pretreatment liquids were analyzed to identify the reactions occurring in the liquid during pretreatment. In NaOH treatment, substantial fraction of dissolved hemicellulose reacts with lignin forming a lignin-carbohydrate complex (LCC). In ammonia treatment, partial hydrolysis of hemicellulose occurs. The carbohydrates in the liquid were primarily xylose oligomers.

603d Hydrolysis of Sucrose in Dilute Sulfuric Acid Solutions at Pretreatment Conditions for Lignocellulosic Biomass

Dan Schell, National Renewable Energy Laboratory, National Bioenergy Center, 1617 Cole Blvd., Golden, CO 80401 and **Ranil Wickramasinghe**, Colorado State University, Department of Chemical Engineering, Fort Collins, CO 80523-1370.

Biomass represents an abundant renewable resource for the production of biobased products and biofuels. Lignocellulosic biomass consists of three main polymers: cellulose, hemicellulose and lignin. The bioconversion of lignocellulosic biomass requires two main processes: cellulose and hemicellulose conversion (hydrolysis) to monosaccharides, followed by fermentation of the monosaccharides into the desired product(s). In addition, lignocellulosic biomass sources such as agricultural residues (corn stover) and herbaceous materials (switchgrass) may contain appreciable amounts of sucrose. The aim of this study was to evaluate the amount of sucrose and its hydrolysis products, fructose and glucose, that remain after dilute sulfuric acid processing at reaction conditions typically used to pretreat lignocellulosic biomass. Solutions containing 25 g/L sucrose with 0.1% to 2.0% (w/w) sulfuric acid concentrations were treated at temperatures of 160°C to 200°C for 3 to 12 min. Sucrose was observed to completely hydrolyze at all treatment conditions. However, appreciable concentrations of fructose and glucose were detected and glucose was found to be significantly more stable than fructose. Different mathematical approaches were used to fit the kinetic parameters for acid-catalyzed thermal degradation of these sugars. Since both sugars may survive dilute acid pretreatment, they could provide an additional carbon source for production of ethanol and other bio-based products.

603c Kinetics And Modeling Of Xylan Hydrolysis And Xylose Dehydration And Reversion During Acid Pretreatment

Mark R. Nimlos[1], Heidi M. Pilath[1], David K. Johnson[1], Michael E. Himmel[1], Xianghong Qian[2], and Haitao Dong[2]. (1) National Renewable Energy Laboratory, 1617 Cole Boulevard, Golden, CO 80401, (2) Mechanical Engineering, Colorado State University, Mechanical Engineering, 1374 Campus Delivery, Fort Collins, CO 80523

Acid pretreatment of biomass is necessary in many cellulosic ethanol processes. This step hydrolyzes the hemicellulose to form fermentable sugars and exposes cellulose for subsequent enzymatic hydrolysis steps. A key to the economic success of cellulosic ethanol production is the efficient hydrolysis of xylans into the monomer, xylose. The yield of xylose is reduced by dehydration reactions, which form furfural; and reversion reactions, which are the reverse of hydrolysis reactions. Reversion reactions occur in high concentration xylose solutions (~200 g/L) and result in formation of xylooligomers. In this study, experimental measurements were used in conjunction with quantum mechanical molecular modeling to explore the fate of xylan, xylooligomers, and xylose in acidic solutions. Experiments were conducted using highly agitated microwave-heated reactor vessels. This reactor set up allowed rapid heating of the sample and limited the effects of mass transport, thus permitting the measurement of intrinsic kinetics. The experimental kinetics and activation energies will be compared to the results of quantum mechanical calculations. We used both static electronic structure theory and quantum mechanical molecular dynamics simulations to obtain activation energies and free energies for these important reactions.

603b Synergistic Hydrolysis Of Afex Treated Corn Stover & Poplar Using Automated 96-Well Microplate Assay

Shishir Chundawat, **Venkatesh Balan**, and Bruce E. Dale. Chemical Engineering and Materials Science, Michigan State University, 3246 Engineering Building, East Lansing, MI 48824

Past research on biomass conversion has focused on acid pretreatment technologies that remove lignin and hemicellulose in the pretreatment step itself. With increasing focus on Ammonia Fiber Expansion (AFEX) pretreatment the need to optimize enzyme systems that hydrolyze cellulose and hemicellulose becomes critical. Modifications in the ammonia recovery and pretreatment reactor systems that utilize low concentrations of aqueous ammonia allow significant cost reductions. However, with reducing severity of pretreatment it becomes important to re-optimize the enzyme cocktail to hydrolyze the pretreated biomass.

The automated 96-well microplate method is a rapid hydrolytic assay technique (scaled down version of NREL LAP-009 protocol) developed at our laboratory to screen optimum enzyme cocktails for hydrolyzing lignocellulosic biomass. Limited work in the past has been directed towards understanding the nature of synergy between cellulases, hemicellulases and other accessory enzymes on lignocellulosic substrates. Recent research has shown that choice of an appropriate enzyme preparation depends more on the characteristics of the substrate rather than on standard enzyme-activities measured.

We have screened commercially available cellulases, xylanases and other accessory enzymes under a wide range of total enzyme concentrations for the hydrolysis of AFEX pretreated corn stover and poplar.

Web Page: www.everythingbiomass.org

603e Quantitative Solid-State NMR Characterization of Solid Residues Produced Using Different Acids in the Hydrolysis of Loblolly Pine

Carsten Sievers, Teresita Marzialetti, Travis Hoskins, Mariefel B. Valenzuela, Pradeep K. Agrawal, and Christopher W. Jones. School of Chemical and Biomolecular Engineering, Georgia Institute of Technology, 311 Ferst Drive, N.W., Atlanta, GA 30332-0100

The production of fuels and chemicals from biomass in liquid phase processes requires depolymerization of the polymeric feedstock. One of the most common approaches for this is hydrolysis in aqueous phase using acid or base catalysts. In the present work different acids (H_2SO_4, H_3PO_4, HNO_3, HCl, TFA) are compared as homogeneous catalysts for biomass hydrolysis. The influence of the proton concentration and the coordination of the corresponding anions to the biomass are investigated. The dissolution yield increased strongly with increasing temperature. In contrast, the effect of the acid concentration was surprisingly small.

A key focus of this work is the characterization of the non-hydrolyzed solid residue by ^{13}C CP MAS NMR spectroscopy. The use of physical mixtures of cellulose and lignin as calibration standards allowed for quantification of the carbohydrate and lignin fractions in the solid residue. The quantitative spectroscopic results are confirmed by proximate analysis, which includes the determination of the lignin and saccharide content by the Klason method and HPLC after complete hydrolysis, respectively. In addition to being a fast and easy analysis method for the determination of the carbohydrate and lignin contents in biomass residues, ^{13}C CP MAS NMR provides information about chemical modifications of the residues. It was shown that the formation of aliphatic side products becomes a significant reaction pathway at 473 K. The results are suggested to be quite useful for optimizing the process conditions for acid hydrolysis of woody biomass.

The aqueous phase was analyzed by LC. As expected, high concentrations of sugars were found after all reactions. The extent of degradation increased strongly with increasing temperature and increasing acid concentration (decreasing pH of the aqueous phase). The most important degradation pathway, yielding furfural and 5-hydroxymethyl furfural (HMF), will be discussed.

603f Differences In The Dilute Acid Pretreatment Of Corn Stover Anatomical Fractions

David K. Johnson[1], David W. Templeton[1], Todd B. Vinzant[1], and Steven R. Thomas[2]. (1) National Renewable Energy Laboratory, 1617 Cole Boulevard, Golden, CO 80401, (2) Ceres, Inc.,, 1535 Rancho Conejo Blvd., Thousand Oaks, CA 91320

A key barrier to the commercialization of fuels and chemicals produced from lignocellulosic biomass is the high cost and relative inefficiency of converting biomass into fermentable sugars. Dilute acid pretreatment is a promising technology for increasing the enzymatic digestibility of lignocellulosic biomass. A better understanding of the pretreatability of biomass is needed so that the rate and yields of sugars can be increased. Corn stover has been identified as a good feedstock for conversion into sugars and ethanol, however, different parts of the corn plant are physically, chemically and ultrastructurally distinct from one another, and so can be expected to respond differently to dilute acid pretreatment and enzymatic saccharification reaction conditions. We have characterized the changes in plant cell wall chemistry that occurred in selected anatomical fractions of corn stover during pretreatment over a range of severities. The fractions include nodes, internode stems, leaves, cobs and sheaths. The release of sugars from dilute acid pretreatment of corn stover anatomical fractions will be described and related to the gross chemical composition of the fractions. Ethanol yields from simultaneous saccharification and fermentation of selected samples will also be reported. Impacts on the yields of sugars from whole corn stover will be discussed and causes for the differences in behavior of the fractions considered.

634a I-Farm - Modeling Biomass Production From A Farmer's Perspective

Ed Van Ouwerkerk[1], **Tom L. Richard**[2], and Rob Anex[1]. (1) Agricultural and Biosystems Engineering, Iowa State University, 3112 NSRIC Building, Ames, IA 50011-3020, (2) Agricultural and Biological Engineering, The Pennsylvania State University, 225 Agricultural Engineering Building, University Park, PA 16802-1909

I-FARM is a web-based farming decision tool designed to simulate whole farm interactions among crops, livestock, carbon and nutrient flows, soil erosion, labor and energy requirements, and overall farm profitability. Recently I-FARM has been expanded to include bioenergy crops and residues, including switchgrass and corn stover. Once farmers have loaded up the biophysical properties of their farm (currently facilitated by a GIS data-mining tool for Iowa) they can evaluate a wide range of scenarios for different crop rotations, yield, price, etc. Results for several farming systems will be presented, demonstrating the variation in price and productivity even in highly productive farmland. Implications for biomass price and policy incentives will be discussed.

634b Effects Of Nitrogen Fertilizer On The Ethanol Fuel System

Seungdo Kim and **Bruce Dale**. Department of Chemical Engineering, Michigan State University, Room 2527 Engineering Building, East Lansing, MI 48824-1226

Ethanol is used as a substitute transportation fuel for gasoline in two ways: E10 (a mixture of 10 % ethanol and 90 % gasoline by volume) and E85 (a mixture of 85 % ethanol and 15 % gasoline by volume). Most ethanol in the United States is derived from corn grain via dry milling or wet milling, and its annual production capacity is about 22 million m3. More ethanol plants, which in total would add up to another 24 million m3 yr-1, are under construction. Using ethanol derived from corn grain as liquid fuel would reduce nonrenewable energy and greenhouse gas emissions, but would increase acidification, eutrophication and photochemical smog, compared to using gasoline as liquid fuel, if current practices remain unchanged (primarily agricultural practices). The dominant factor influencing greenhouse gas emissions, acidification, eutrophication and photochemical smog formation is soil related nitrogen losses (e.g., N2O, NOx and NO3-). Nitrogen fertilizer applied during corn cultivation is probably the primary source of soil related nitrogen losses. A higher application rate of nitrogen fertilizer would produce more corn grain. However, more soil related nitrogen emissions would also be produced at a higher application rate of nitrogen fertilizer. The agricultural processes also account for 27 – 44 % of the total energy consumption in producing biobased products. This study investigates the effects of nitrogen fertilizer application on the bioethanol fuel system. To estimate local soil carbon and nitrogen dynamics in biomass cultivation, this study uses the DAYCENT model, which is the daily time step version of the CENTURY model, a multicompartmental ecosystem model. The system boundary includes corn production, transportation of corn to a dry mill, the dry mill processing steps, transportation of ethanol to users, ethanol fueled vehicle operation, and upstream processes. The allocation procedures for external functions are done by the system expansion approach. To compare the environmental performance between bioethanol fuel and gasoline fuel, gasoline fueled vehicle operation is included in the system boundary. Planting winter cover crops after harvest is one possible way to reduce nitrogen losses from soil because winter cover crops serve as a nitrogen scavenger. We also investigate the effects of planting winter cover crops on the bioethanol fuel system. The potential impact categories considered here are nonrenewable energy consumption, greenhouse gas emissions, photochemical smog, acidification, and eutrophication. Environmental impacts are estimated by characterization factors given by the United States Environmental Protection Agency (EPA-TRACI).

634c Optimal Utilization Of Biomass Inputs In An Integrated Biorefinery

Norman E. Sammons Jr.[1], Wei Yuan[1], Mario R. Eden[1], Burak Aksoy[2], and Harry T. Cullinan[2]. (1) Department of Chemical Engineering, Auburn University, 132 Ross Hall, Auburn University, AL 36849-5127, (2) Alabama Center for Paper and Bioresource Engineering, 242 Ross Hall, Auburn University, AL 36849

The integrated biorefinery, which uses renewable feedstocks from the forest-based industries, has the opportunity to provide a strong, self-dependent, sustainable alternative for the production of bulk and fine chemicals from polymers, fiber composites and pharmaceuticals to energy, liquid fuels and hydrogen. With such a wide range of processing steps and possible products, it is obvious that identification of the optimum process structure can not be done based on heuristics or rules of thumb. Changing market conditions may dictate a dynamic optimum for the allocation of resources and production capacity, resulting in a myriad of possible long-term product portfolios. Economic market analysis, predictive financial modeling, and optimization under uncertainty are tools that can be used to determine the sensitivity of a decision-making framework to market fluctuations. Thus there is a need for systematic, reliable methods capable of incorporating different levels of process detail in the decision making framework. In this work a mathematical optimization based framework is being developed, which enables the inclusion of profitability measures and other techno-economic metrics along with process insights and performance characteristics obtained from experimental and modeling studies. By utilizing process integration methods, the processing steps can be optimized to ensure efficient use of energy and materials resources while assuring an acceptable, minimal level of environmental impact through the use of EPA's WAR algorithm. An inherent benefit of the proposed framework comes from the decoupling of the complex models from the selection step, which results in the ability to adapt to new developments within any of the processing steps and thus also incorporate novel innovative production processes in the decision-making framework. In this way, experimental and theoretical efforts can supplement each other in a synergistic manner, by providing direction and data for continued work. This contribution will illustrate the strategy for developing the decision making framework as well as highlighting the flexibility of the framework to utilize data from technological breakthroughs in the field of biorefining.

634d Life Cycle Analysis Of A Renewable Lumber Preservative

Monty Singletary[1], Mark Bricka[1], and Phile Steele[2]. (1) Dave C. Swalm School of Chemical Engineering, Mississippi State University, Mississippi State, MS 39762, (2) Forest Products Department, Mississippi State University, Mailstop 9820, Mississippi State, MS 39762

The Chemical Engineering Department, in conjunction with the Agricultural and Biological Engineering Departments at Mississippi State University, has designed a pyrolysis process to convert woody biomass into a "Bio-Oil" for use as a preservative in the lumber treatment industry. Historically, CCA (a mixture of copper, chromium, and arsenic) has been used as a wood preservative, but it has obvious toxic characteristics and its use has been banned in the US. Our Bio-Oil has shown promise as a natural alternative to this preservative. In preparation for full-scale production, a cradle-to-grave analysis has been performed on the lab-scale pyrolysis unit, involving feedstock and construction materials as well as storage and disposal issues. This paper presents the potential environmental and economic impacts of this process through Life Cycle Analysis as guided by NREL and EPA.

LIFE CYCLE ANALYSIS OF CHARCOAL PRODUCTION FROM BIOMASS

Dr. Sudhagar Mani, University of Georgia, Athens, GA
Dr. K.C. Das, University of Georgia, Athens, GA

Extended Abstract

Charcoal has been traditionally produced from wood and used as a primary biomass fuel for cooking and heating in most of the developing countries. Production of charcoal from traditional methods is inefficient and demands critical amount of energy and generates pollution and land degradation problems. Despite its inefficient production, charcoal has been recognized for its high heating value and rich source of carbon for industrial applications. In addition, charcoal has been applied to soil for carbon sequestration and acts as a growth promoter for certain crops. Charcoal has also been studied as a potential candidate for gasification process to produce high C/H ratio syngas for biofuel production and used as an absorbent to remove toxic volatile organic compounds. Recently, significant progress has been made to improve the efficiency of charcoal production with the maximum solid yield of up to 60%. In order to develop sustainable production of charcoal from biomass with minimal environmental impacts, a life cycle analysis (LCA) tool was used to quantify energy use and environmental emissions from various production methods and was compared. The main objective of this study was to conduct a cradle-to-grave analysis for producing charcoal from forestry sources into its final utilization and disposal (Figure 1). Life cycle inventory data for this analysis was taken from different literature sources including SIMAPRO – databases and Environmental Protection Agency (EPA). In this study, three different charcoal making processes were investigated (Figure 2): 1) traditional kiln type (eg. Missouri kiln, direct heat pyrolysis), 2) Industrial charcoal reactors (eg. Linann reactor, indirect heat pyrolysis) and 3) Continuous pyrolysis reactor (eg. UGA pyrolysis unit, indirect heating pyrolysis process). In the Linann charcoal process, the liquid co-product called vinegar was produced and used for many different applications in the developing countries. The University of Georgia (UGA) developed a continuous pyrolysis process at medium temperature conditions to successfully produce both charcoal and bio-oil from various biomass sources. For the both indirect heating pyrolysis system, a mass-based resource (energy, fuel, emissions) allocation procedure was used to estimate the energy use and emissions to produce charcoal. A functional unit of any resources/t of charcoal produced was used. The potential impact categories considered in this analysis are energy use, climate change, smog formation, acid rain formation and human health impacts. Development of a sustainable charcoal production method can help reduce energy use, environmental emissions, land degradation and worldwide deforestation, while producing high quality charcoal for various industrial applications.

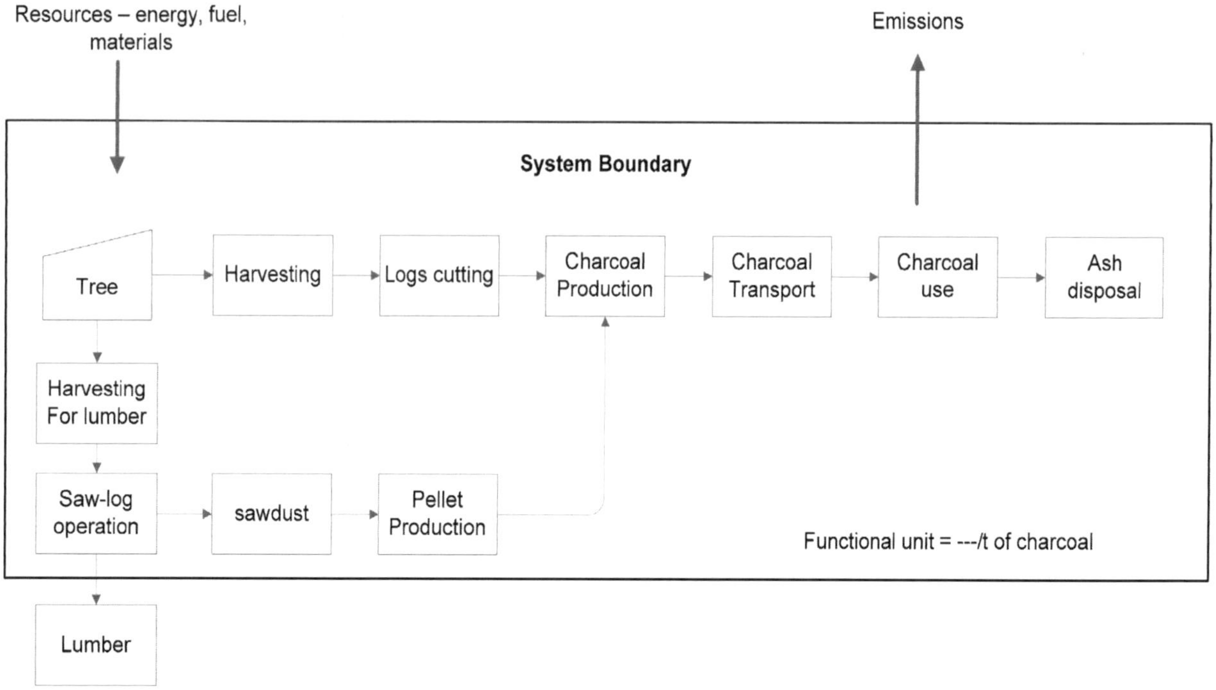

Figure 1. Schematic flow diagram of producing charcoal from forest sources

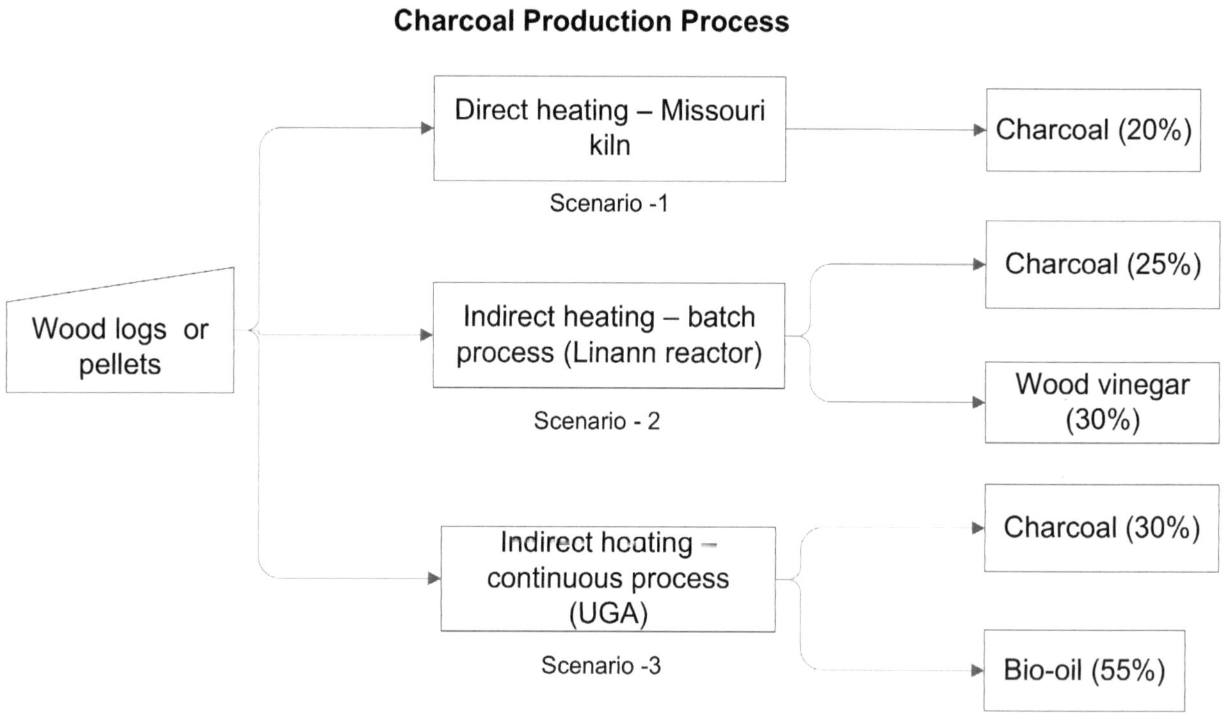

Figure 2. Schematics of different charcoal making processes

634f An Integrated Systems Approach To Reducing Energy Inputs While Maximizing Conversion To Ethanol From Alternative Feedstocks

Kimberly Ogden, University of Arizona, Department of Chemical and Environmental Engineering, P. O. Box 210011, Tucson, AZ 85721

Alternative fuel production is of increasing interest throughout the U.S. and the world as cost and demand increase while fossil fuel supplies decrease on a global scale. The state of Arizona presents a unique situation as we have an abundance of sunny days which provide an ideal environment for the growth of plants that could be utilized as feedstocks for production of fuels. The goal is to design a system to primarily produce ethanol from biorenewable feedstocks in a sustainable manner that requires low inputs of nutrients, energy, and water. An ideal system for Arizona integrates an appropriate adapted feedstock grown with recycled or reclaimed water. Agriculture in Arizona is abandoning traditional cotton growth due to recent economic changes and so there are facilities across the state which have available land and available allocated water resources. Working collaboratively, we are investigating optimization of ethanol production from sweet sorghum and/or algae. Sweet sorghum is chosen as a feedstock since it is salt tolerant and requires less seed, fertilizer, pesticide, irrigation water, and tillage than other crops currently used for ethanol production. CO2 utilized by algae in photo-bioreactors is the second system investigated because it can produce carbohydrates year-round that are used as feed for the fermentation vessel and use recycled/reclaimed water. Both systems however have limitations. Sweet sorghum has a high sucrose content in addition to fructose and glucose which presents fermentation challenges. Furthermore, there are large quantities of cellulosic material and transportation of this material may be too costly. Fermentation results and a preliminary optimization strategy for balancing production costs will be presented.

644a Conformational Energies of Cellobiohydrolase I Linker on Cellulose Surface

Xiongce Zhao, Center for Nanophase Materials Sciences, Oak Ridge National Laboratory, Oak Ridge, TN 37831, Clare McCabe, Department of Chemical Engineering, Vanderbilt University, Nashville, TN 37235-1604, and Michael E. Himmel, National Bioenergy Center, National Renewable Energy Laboratory, 1617 Cole Blvd., Golden, CO 80401.

Cellobiohydrolase I (CBH 1) from Trichoderma reesei is one of the most active enzymes in the hydrolyzation of cellulose. CBH 1 is a multi-domain enzyme, consisting of a large catalytic domain containing an active site tunnel and a small cellulose binding module joined to one another by a 27 residue linker peptide, that is believed to hydrolyze cellulose in a "processive" manner. However, the exact mechanism of the depolymerization of cellulose by CBH I is not fully understood. In this work we have performed molecular dynamics simulations that focus on determining the role and function of the linker peptide. Although the sequence of the linker peptide is known and it has been shown that the linker plays an important role in enzymatic activity during cellulose hydrolysis, the spatial conformation adopted by the linker domain is yet to be determined. We probe this through molecular dynamics simulations in which the motion and conformation of the linker adsorbed on a cellulose surface in an aqueous environment are monitored. Potential of mean force calculations have been performed to compute the free energy of the linker as a function of linker length. We compare our simulation data with the hypothesis that the linker behaves like a spring that helps the enzyme to move along the cellulose chain. The goal of the simulations is to help elucidate the fundamental relationship between the structure of the linker and catalysis process of the enzyme, and to provide insight into designing more effective enzymes for the conversion of biomass.

644b Kinetic Measurements Of The Enzymatic Hydrolysis Of Microcrystalline Cellulose By In Situ Afm

Dean C. Dibble, Energy Systems, Sandia National Laboratories, P.O. Box 969, MS 9161, Livermore, CT 94551 and Blake A. Simmons, Sandia National Laboratory, 7011 East Avenue, Livermore, CA 94550.

An important challenge for the large-scale production of cellulosic ethanol is the development of efficient cellulase enzymes on an industrial scale. However, specific details of the interaction between cellulase enzymes and their substrates are not yet understood. Although the structure of cellulase from several organisms has been determined and provides some insight into the process of saccharification, there is a need for experimental measurements to explain macroscopic observations such as the resistance to cellulase activity of some fractions of cellulosic biomass and to understand the properties of natural cellulose that influence the enzymatic hydrolysis process. In this talk we describe our measurements of the enzymatic hydrolysis process obtained in situ, and with high spatial resolution. The microcrystalline cellulose we have chosen as a substrate was derived from wood pulp and is expected to be similar to the crystalline components of cellulose from biomass sources that are candidates for ethanol production. Sequences of atomic force microscope (AFM) images at a single location and spaced at regular time intervals enabled measurements of hydrolysis rates on individual cellulose microfibrils both individually and when part of a fibril bundle as a function of time. Our work appears to be the first to describe the time resolved interaction of cellulose with cellulase at high spatial resolution.

644c Pretreatment Of Biomass With Surfactants

Qing Qing, Michael Studer, **Bin Yang**, and Charles Wyman. University of California Riverside, 1084 Columbia Avenue, Riverside, CA 92507

Lignin in pretreated cellulosic biomass can non-productively adsorb cellulase, resulting in loss of a significant portion of this expensive protein. In addition, lignin interferes with the path for cellulase action, slowing down hydrolysis. Thus, the effectiveness of enzymatic hydrolysis of pretreated lignocellulosic biomass can be significantly enhanced if lignin is removed or modified before adding enzymes. In this project, the effect of adding various surfactants as well as various surfactant concentration were evaluated for the enzymatic digestibility of corn stover pretreated at 140-220℃ with and without sulfuric acid. In addition, cellulase adsorption on pretreated solids was significantly reduced, apparently as a result of the interaction of the surfactants with lignin. A mechanism will be presented to explain the observed impact of surfactants on performance.

644d Microbial Conditioning and Pretreatment of Grasses for Ethanol and Lignin Co-Products

Swetha Mahalaxmi[1], Ashwini Thakre[1], Clint Williford[1], and Colin Jackson[2]. (1) Chemical Engineering, University of Mississippi, 134 Anderson Hall, Post Office Box 1848, University, MS 38677, (2) Department of Biology, University of Mississippi, 134 Anderson Hall, Post Office Box 1848, University, MS 38677

Cellulosic ethanol continues to develop as a potential alternative transportation fuel. For its production, major process steps involve pretreatment to degrade or disengage lignin, saccharification to depolymerize cellulose to sugars, and fermentation of sugars to ethanol. Through DOE-supported projects, Novozymes and Genencor have dramatically reduced the enzymatic hydrolysis costs, leaving pretreatment as a significant remaining cost challenge to cellulosic ethanol commercialization. Aggressive research and development is pursuing alternative pretreatments to meet this challenge. In this matter, lignin plays a key role. It inhibits enzymatic hydrolysis of cellulose, and process designs typically consign its recalcitrant, low-value forms to boilers. However, lignin, in appropriate molecular weight ranges, is a suitable component of phenolic resin plastics.

Our objective is to combine microbial conditioning to improve pretreatment performance, allowing reductions in severity and complexity, reduced costs, and improved economics for the biomass grower. Furthermore, we aim to obtain a more useable lignin co-product.

In this work, we applied a wastewater sludge bacterial (consortium) and Phenerochaete Chrysosporium to Corn Stover, Switch Grass, Sorghum, and Giant Miscanthus for 20 and 28 days respectively. We measured enzyme activities for Phenol oxidase, Peroxidase and beta-Glucosidase during the period of fungal conditioning of the biomasses. The bacterial consortium was characterized for the active species responsible for conditioning. The conditioned biomasses were washed and analyzed for their composition. Sugar yields after 72 hours of enzymatic hydrolysis were determined using HPLC. Finally, a comparison of the impact of Ammonia Fiber Expansion (AFEX), Extrusion and Centrifugation, Dilute Acid Hydrolysis and Microbial Conditioning on enzymatic hydrolysis was made on the four feed stocks. The liquid portion of the conditioned material was analyzed for solubilized sugars and lignin. Findings include observations that bacterial conditioning liberated significant xylose and that fungal conditioning generated primarily the beta-Glucosidase enzyme.

644e Interactions Of Ensiled Storage With Biomass Pretreatment

Tom L. Richard[1], Qin Chen[1], Corey Radtke[2], Lynn Petzke[2], Megan Marshall[1], and Kevin Shinners[3]. (1) Agricultural and Biological Engineering, The Pennsylvania State University, 249 Agricultural Engineering Building, University Park, PA 16802-1909, (2) Biotechnology, Idaho National Laboratory, P.O. Box 1625, Idaho Falls, ID 83415, (3) Biological Systems Engineering, University of Wisconsin, 460 Henry Mall, Madison, WI 53706

Recent evaluations by the DOE and others indicate that large quantities of biomass grown in semi-humid and humid regions are likely to be harvested and stored wet rathert than in dry bales. Under anaerobic conditions wet storage frequently produces sufficient organic acids to limit biodegradation by reducing pH, resulting in significantly lower mass losses than in dry outdoor storage. Similar acid fermentations, commonly called ensilage, have been used for biomass storage in livestock production systems for over a millennium. This study investigates the interaction of ensiled storage with reduced severity dilute acid pretreatment, and examines the resulting impact on hydrolysis and saccharification. Results indicate ensiled storage creates a favorable environment for microbial and enzymatic biocatalysis, reducing required pretreatment severity, and subsequently increasing cellulose and hemicellulose hydrolysis to sugars for ethanol fermentation. Chemical and thermal pretreatment techniques currently constitute over 30% of the costs of biomass conversion to ethanol, so integration of beneficial wet storage strategies with pretreatment could significantly increase the economic viability of cellulosic biorefineries.

644f Modified Dry Grind Process Integrating Cellulosic Ethanol From Distiller's Grains

Youngmi Kim[1], Rick Hendrickson[1], Nathan Mosier[2], and Michael Ladisch[3]. (1) Laboratory of Renewable Resources Engineering (LORRE), Purdue University, Potter Engineering Center, 500 Central Drive, West Lafayette, IN 47907-2022, (2) Agricultural and Biological Engineering and Laboratory of Renewable Resources Engineering (LORRE), Purdue University, Potter Engineering Center, 500 Central Drive, West Lafayette, IN 47907-2022, (3) Agricultural and Biological Engineering & Biomedical Engineering, and Laboratory of Renewable Resources Engineering (LORRE), Purdue University, Potter Engineering Center, 500 Central Drive, West Lafayette, IN 47907-2022

Faced with an escalating demand for fuel ethanol, researchers and ethanol industries are searching for ways to enhance ethanol yield by incorporating cellulosic ethanol technologies into current ethanol processes. Distiller's grains are a major co-product of dry grind ethanol industry that is rich, not only in proteins, but also in unutilized polymeric sugars, such as cellulose, xylan and residual starch. Efforts are underway to utilize and process co-products from dry grind ethanol processes for additional fermentable sugars and ethanol, while preserving the value of the residue as an animal feed. Efficient digestibility of distiller's grains was demonstrated from a laboratory scale hydrolysis of distiller's dried grains, with solubles (DDGS) pretreated by aqueous hot liquid water. This resulted in 92% conversion of total glucan to glucose within 24 hours by 15 FPU of cellulase per gram of glucan. Including a complete compositional analysis of distiller's grains, its digestibility and fermentability. This paper reports a model of a proposed dry grind ethanol process that integrates cellulosic ethanol technologies for enhanced ethanol yield through use of distiller's grains. Theoretical ethanol yields, compositions of resulting co-products, and challenges in implementing a modified dry grind process are discussed.

655a Optimization of a Biomass Fractionation Process and Evaluation of Its Economic Feasibility

Kiran L. Kadam, Chim Y. Chin, and Lawrence W. Brown. PureVision Technology, 511 N. McKinley Ave., Fort Lupton, CO 80621

PureVision Technology, Inc.'s two-stage process fractionates biomass into three streams: a solid cellulose stream and two liquid streams containing mostly hemicellulosic sugars and lignin, respectively. Process optimization was initiated for both stages using corn stover as feedstock in a continuous process development unit (PDU) with the goals of efficient hemicellulosic and lignin hydrolysis for the first and second stage, respectively. Both autohydrolysis and acid-catalyzed prehydrolysis were evaluated for the first stage, and autohydrolysis was selected as the preferred mode for the first stage. Although autohydrolysis achieves lower xylose recovery (about 65% at 210C) compared to acid-catalyzed pretreatment, it also offers benefits including: 1) no cost of acid, 2) cheaper materials of construction, 3) no gypsum formation and related disposal and handling costs, and 4) no need for L/S separation after liquor conditioning. For the second stage, 220C was determined to be optimal with NaOH charge being 0.06 g/g biomass, which is half of that in the kraft process. Using the level of performance currently possible using the PDU, economic feasibility of the PureVision process for ethanol production was compared to that of the NREL process. Major differences between the two scenarios are that the PureVision process generates lignin as a sellable output but needs to replace the lost fuel value of the lignin with additional stover. An average lignin selling price of $330/mt is used based on lower end applications such as concrete binder ($250/short ton) and feed binder ($350–425/short ton). The minimum ethanol selling price (MESP), calculated following the NREL protocol, is projected to be lower for the PureVision process than that for the NREL process, the differential depending on lignin selling price. Potential for further lowering the MESP exists by reducing NaOH usage by tweaking the second stage parameters, and by reducing corn stover consumption for fuel by fully optimizing the turbogenerator operation. As these changes are within practical realm, PureVision's process is economically attractive and has excellent commercial potential.

Pretreatment and Fractionation of Corn Stover by S.E.A.A. (Soaking In Ethanol and Aqueous Ammonia)

Tae Hyun Kim, Department of Agricultural and Biosystems Engineering, Iowa State University, Ames, IA
Nhuan-John Nghiem, Kevin B. Hicks, USDA-ARS-ERRC, Wyndmoor, PA

Introduction

Soaking in Aqueous Ammonia (SAA) has been proven to be an effective pretreatment method for lignocellulosic biomass (1, 2). SAA is only one of the few methods for pretreatment of biomass in which almost 100% glucan and greater than 80% xylan are retained whereas a significant percentage of lignin is removed. Despite this high efficiency in carbohydrate preservation still about 20% xylan is solubilized together with lignin and hence is not available for conversion to ethanol in the subsequent fermentation step.

In this study, we have investigated a modification of the SAA method, a Soaking in Ethanol and Aqueous Ammonia method (SEAA), to reduce the loss of xylan due to solubilization. In this method ethanol is added to the aqueous ammonia solution. Precipitation of hemicellulose with ethanol has been investigated for the separation and purification of polysaccharides in the biomass (3, 4, 5). The presence of ethanol causes re-precipitation of the previously solubilized xylan onto the solid matrix. Thus more xylan is available for conversion to ethanol. The ethanol used in the pretreatment process can be recovered and recycled. In addition to increased availability of the carbohydrates for ethanol production, the solubilized lignin theoretically can be recovered in relatively pure form and converted to high-value products. Re-precipitation of the solubilized xylan does not affect the bioconversion efficiency of the pretreated biomass. The pretreated biomass can be converted to ethanol or chemicals in the simultaneous saccharification and co-fermentation (SSCF).

Materials and Methods

Corn Stover

Corn stover was supplied by the National Renewable Energy Laboratory (NREL, Golden, CO). It was ground and screened. The fraction collected between of 10–35 mesh (0.5-2.0 mm) was used in all experiments. It was air dried and stored in the ambient conditions. Moisture content of ground corn stover was approximately from 5% to 8%. Initial compositions of corn stover are shown in Table 1.

Table 1

Initial composition analysis of untreated corn stover

Description	Composition [wt. %]
Glucan	36.7 ±0.2 [2]
Xylan	22.1 ±0.2
Galactan	1.1 ±0.0
Arabinan	3.8 ±0.0
Mannan	0.3 ±0.0
Acid in-soluble lignin	14.4 ±0.1
Acid soluble lignin	N/A
Ash	6.1 ±0.1

Note. 1. Data in the table are based on the oven dry untreated biomass

2. Standard deviation (n=2).

Enzymes

Cellulase enzyme, GC-220 (Genencor International Inc., Lot #301-04232-162), was obtained from Genencor International. The average activity of the enzyme was 45 filter paper unit (FPU)/mL and the protein content was 184 mg/mL. Activity of β-glucosidase (Novozyme 188 from Novo Inc., Lot No. 11K1088) was 750 CBU/mL.

Digestibility Tests

The enzymatic digestibility of barley hull was determined in duplicate according to the NREL Chemical Analysis and Testing Standard Procedure (NREL, 2004). The conditions of the enzymatic digestibility tests were 50°C and pH 4.8 (0.05 M sodium citrate buffer) on a shaker bath agitated at 150 rpm. Enzyme loadings of 15 FPU of GC-220/g-glucan supplemented with 30 CBU of β-glucosidase (Novozyme 188)/g-glucan were used. The initial glucan concentration was 1% (w/v) based on 100 mL of total liquid and solid. All the samples used in the digestibility tests were wet samples as collected from various pretreatments. The 250 mL screw-capped Erlenmeyer flasks containing the enzyme hydrolysis preparations were placed in an incubator shaker (Lab-line, 4827F, Dubuque, Iowa). Samples were taken periodically (6, 12, 24, 48, 72 and 96 h) and analyzed for glucose, xylose, and cellobiose content using HPLC (High Performance Liquid Chromatography) (6). Total released glucose (or xylose) after 72 h of hydrolysis was used to calculate the enzymatic digestibility. Avicel® PH-101 and untreated barley hull were put through the same procedure as a reference and control, respectively.

Analytical Method

The solid samples, such as treated/untreated barley hull, Avicel® PH-101, etc., were analyzed for sugar and Klason lignin following NREL Chemical Analysis and Testing Standard Procedures (NREL, 2004). Each sample was run in duplicate. Sugars were determined by HPLC using a Bio-Rad Aminex HPX-87P column and a refractive index detector (6).

An YSI 2300 Glucose/Lactate analyzer (YSI Incorporated, Yellow Springs, Ohio) was also used for rapid analysis of glucose. A refractive index detector was used for HPLC analysis.

Acid insoluble lignin and carbohydrate content were determined by NREL Chemical Analysis and Testing Standard Procedure.

Experimental Setup

The convection oven and SS-316 column reactor (shown in Figure 1) were used. In a SEAA experiment, 5 g of biomass were packed into the reactor, soaked with ammonia and ethanol solution and put in the pre-set convection oven (60°C) for 24 hours. After the reaction, all the solid and liquid were collected and the filtered. The solid was subject to the carbohydrate and lignin analysis and enzymatic digestibility test (6).

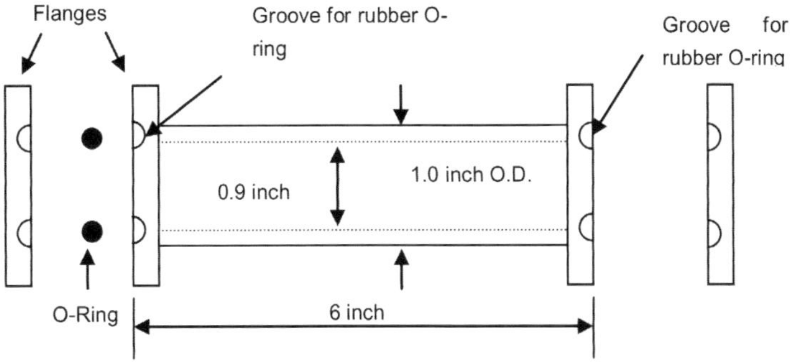

Figure 1. Batch reactor assembly

Results

Ethanol effect on the retention of hemicellulose in solid form.

Table 1 presents the results of the effect of ethanol concentration on the compositions in Soaking in Ethanol and Aqueous Ammonia (SEAA) treatment. In order to observe the ethanol effects on SEAA, total ammonia input were kept in 15 wt. % in total liquid and four different ethanol concentrations (1, 5, 20, and 49 wt. %) were applied, keeping solid:liquid ratio at 1:9 (for example, 5 g corn stover + 45 g total liquid). Ethanol 49% is the maximum ethanol concentration which can be produced in the laboratory since the 30% ammonium hydroxide is the maximum available ammonium hydroxide.

It was observed in Table 1 that xylan fractions, amount of remaining xylan after treatment, were found to be increased by addition of ethanol. Xylan dissolution was 17.2% without ethanol, 16.7% with 1%, 14.5% with 5%, 10.4% with 20%, and 6.3% with 49% ethanol. The glucan content was well preserved showing no significant changes over the entire test conditions. Some unusual behavior in the lignin removal (delignification) was found that the delignification in the case of 1% ethanol was substantially lower than those of 5 % and 20% cases. Delignification was again reduced after 49% ethanol addition. The reason of this unexpected result is unclear at this time.

Table 2. Effect of ethanol concentration on the compositions in Soaking in Ethanol and Aqueous Ammonia (SEAA) treatment [1]

Ethanol concentration	S.R.[2]	Lignin[3]	Delignification	Solid	
				Glucan	Xylan
[wt. %]	[%]	[%]	[%]	[%]	[%]
Untreated	100	14.4	-	36.1±1.0	22.1±0.0
0	65.1±2.5	3.8±0.5	73.6±4.9	35.3±1.0	18.3±0.6
1	65.5±1.1	4.8±0.7	65.3±6.9	35.1±0.8	18.4±0.2
5	66.6±1.5	4.2±0.3	70.8±2.9	35.4±1.4	18.9±0.3
20	68.9±2.4	4.3±0.7	70.1±6.9	35.4±0.5	19.8±0.6
49	72.5±1.8	5.3±0.4	63.2±3.9	35.6±0.6	20.7±0.5

Note. 1. Data in the table are based on the oven dry untreated biomass (n=2). Pretreatment conditions: 15 wt.% of ammonia concentration, 60°C of reaction temperature, and 1:9 of solid:liquid ratio (based on wt.).
2. S.R. stands for solid remaining after reaction.
3. Acid insoluble lignin.

Enzymatic digestibility test

The enzymatic digestibilities of the SEAA-treated corn stover for glucan and xylan are shown in Figure. 2. The pretreatment conditions were 1:9 of solid-to-liquid ratio, 1-49% ethanol, 60°C, 24 h, and 15 wt.% ammonia. With an enzyme loading of 15 FPU/g-glucan, the 1-20% SEAA-treated corn stover showed digestibility in the range of 82-86% of glucan and 67-70% of xylan, whereas the 49% SEAA showed 72% and 58% of glucan and xylan digestibilities, respectively. Increasing the ethanol more than 20% had negative effect on the enzymatic digestibility (Figure. 2). Data in this figure indicate that a significant adverse effect on the digestibility was seen with excessive ethanol. Table 2 and Figure 2 collectively indicate that ethanol decrease the lignin removal by more than 20%, which may affect the enzymatic digestibility of SEAA-treated corn stover.

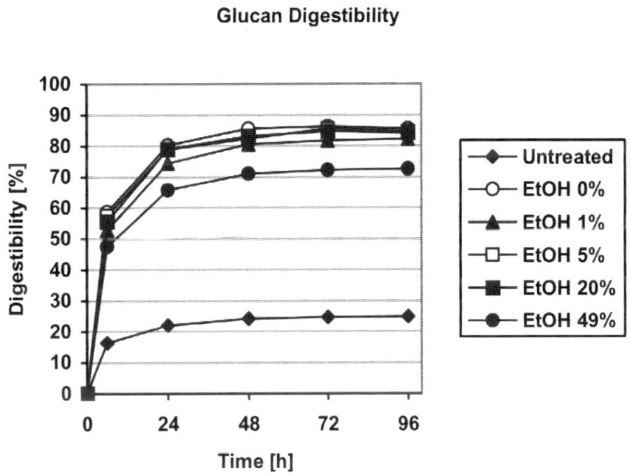

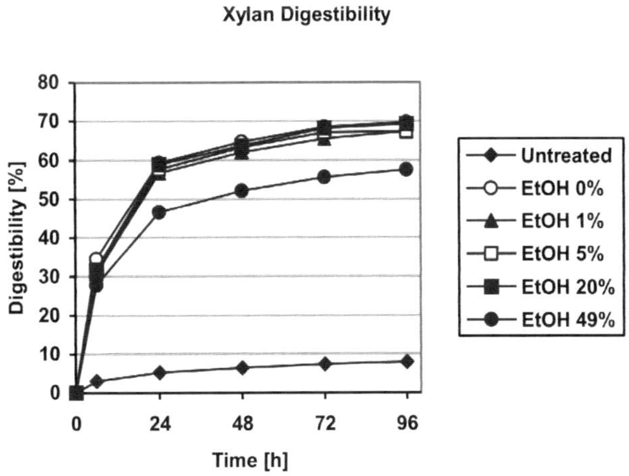

Figure 2. Enzymatic digestibility of SEAA-treated corn stover

Note. (15 FPU cellulase+30 CBU of β-glucosidase)/g glucan enzyme loading

Xylan retention and selectivity

The purpose of this study is to retain the carbohydrates in the solid, which is to produce the carbohydrate-rich solid cake with significantly low lignin content. Carbohydrates, such as glucan, xylan, galactan, arabinan, and mannan, are valuable intermediaries which can be converted into the fuels and chemicals by chemical and biological methods. In the biological conversion of biomass, the lignin has a significant inhibition effect on enzyme activity and microbial reaction (7, 8, 9, 10, 11)

In this study, we introduced a couple of simple definitions as follows:

Xylan retention [%] = $(X_R / X_T) \times 100$

Where,

X_R: Xylan remaining

X_T: Total xylan content in the untreated biomass

Selectivity [-] = m_{Lignin} / m_{Xylan}

Where, m_{Lignin} and m_{Xylan} are the mass loss rate of lignin and xylan from the solid.

Selectivity was calculated and summarized in Figure 3. This result clearly indicates that the selectivities at various conditions increase as ethanol concentration increase. On the basis of results, it can be concluded that the addition of ethanol retains more xylan component in the solid; however, the selectivity was defined as the ratio of lignin removal/xylan removal. It can't be concluded that the higher lignin removal solely occurs at higher selectivity.

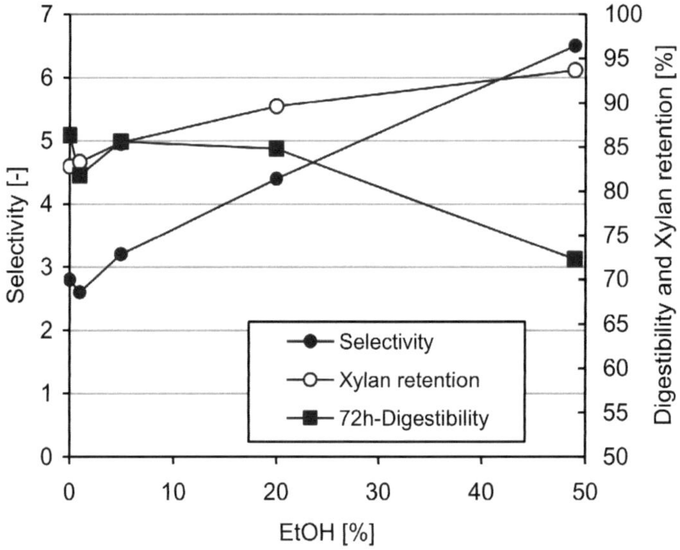

Figure 3. Selectivity plot of the SEAA-treated corn stover upon various ethanol concentrations

Note. 1. Data in the figure are based on the oven dry untreated biomass; Pretreatment conditions: 24 h of reaction time, 1:9 (based on wt.) of solid:liquid ratio, 15 wt.% of ammonia concentration, 60°C of reaction temperature. 2. Selectivity = m_{Lignin} / m_{Xylan} (where, m_{Lignin} and m_{Xylan} are the mass loss rate of lignin and xylan in the solid, respectively)

Xylan retention % upon various ethanol concentrations are summarized in Figure 4. The results obtained by increasing ethanol adding from 0% to 49% shows that xylan fraction undergoes reprecipitation reaction and forms the solid. It is concluded that the ethanol plays an important role in the precipitation of xylan. In Figure 4, insoluble lignin content also increased as ethanol concentration increased; therefore, it was found that both contents increased. At 49% ethanol addition, insoluble lignin content was not significantly changed, but the xylan retention % was gradually increased. However, the 72 h-digestibility of 49% ethanol treated corn stover substantially lower than digestibilities of SEAA-treated corn stovers. The optimum ethanol concentration of 20% for SEAA was therefore found from the standpoint of enzymatic digestibility, selectivity, and xylan retention.

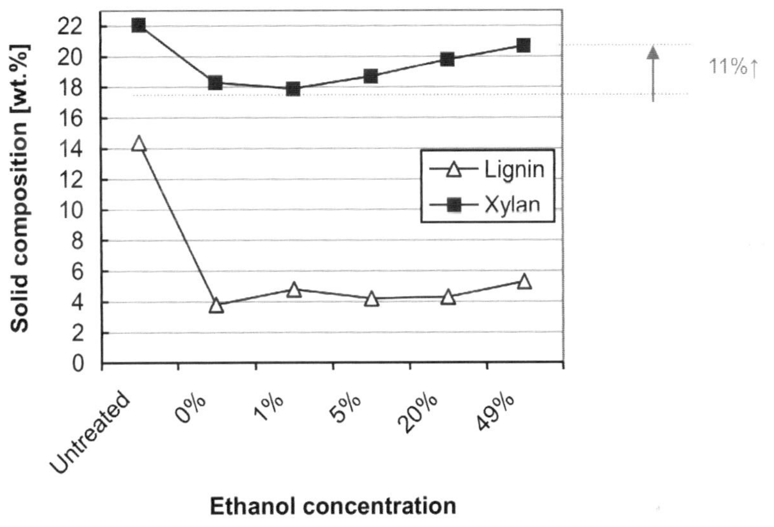

Figure 4. Compositional change in SEAA-treated corn stover with various ethanol concentrations

Conclusion

The addition of ethanol was proven to be effective for the retention of hemicellulose during SEAA treatment of corn stover. With optimum ethanol addition at 20%, the solid sample obtained from SEAA treatment contain 35.4% glucan, 19.8% xylan, and 4.3% lignin, which are almost 100% glucan retention, 89.6% xylan retention, and 70.1% lignin removal, respectively. SEAA was proven to be effective method to produce the carbohydrate-rich solid cake, which can be an intermediate for many chemicals and fuels. Enzymatic digestibility test showed the lignin content in solid may play an important role in the enzymatic hydrolysis of SEAA-treated corn stover.

Acknowledgements

The majority of the support for this research was provided by the USDA-ARS-ERRC in Wyndmoor, PA. Partial support from Department of Agricultural and Biosystems Engineering at Iowa State University in Ames, IA is also acknowledged.

References

1. Kim, T.H. and Lee, Y.Y. (2007), "Pretreatment of Corn Stover by Soaking in Aqueous Ammonia at Moderate Temperature". *Appl. Biochem. Biotechnol.* Vol 136-140, pp 81-92.
2. Kim, T.H. and Lee, Y.Y. (2005), "Pretreatment of Corn Stover by Soaking in Aqueous Ammonia", *Appl. Biochem. Biotechnol.* Vol 124, pp 1119-1132.
3. Whistler, R. L. and Richards, E. L. (1970), "*The Carbohydrates, Chemistry and Biochemistry*"; Pigman, W.; Horton, D., Eds.; Academic Press: New York, Vol. IIA, pp 447–469.
4. Sun, R.C., and Tomkinson, J. (2001), "Fractional separation and Physico-Chemical Analysis of Lignins from the Black Liquor of Oil Palm Trunk Fibre Pulping", *Separation and Purification Technology*, Vol 24, pp 529-539.
5. Anderson, E., Mechtman, J., and Seeley, M. (1938), "Hemicellulose from Cottonseed Hulls", *J. Biol. Chem.,* Vol 126 (1): pp 175-179.
6. NREL (1996), "Chemical Analysis and Testing Laboratory Analytical Procedures (CAT)", National Renewable Energy Laboratory, Golden, CO.
7. Kim, T.H., Kim, J.S., Sunwoo, C., and Lee, Y.Y, (2003), "Pretreatment of Corn Stover by Aqueous Ammonia",. *Bioresource Technology*, Vol 90, pp 39-47.
8. Chang, V.S., and Holtzapple, M.T., (2000). "Fundamental factors affecting biomass enzymatic reactivity", *Appl. Biochem. Biotechnol.* Vol 84/86, pp 5–37.
9. Converse, A.O., (1993). "Substrate factors limiting enzymatic hydrolysis. Biotechnology in Agriculture", *No.9, CABI*, UK, 93–106.
10. Cowling, E.B., and Kirk, T.K., (1976). "Properties of cellulose and lignocellulosic materials as substrates for enzymatic conversion processes". *Biotechnol. Bioeng Symp.* Vol 6, pp 95–123.
11. Dulap, C.E., Thomson, J., and Chiang, L.C., (1976). "Treatment processes to increase cellulose microbial digestibility". *AIChE. Symp., Ser.* Vol 72 (158), p 58.

655c Biomass and MSW to Liquid Fuels and Chemicals Via Gasification

Ben Phillips, Emery Energy, 159 West Pierpont Avenue, Salt Lake City, UT 84101

Emery Energy Company is actively engaged in the design and development of an integrated thermochemical platform for converting biomass and MSW-derived feedstocks into liquid fuels and chemicals. This development effort is currently focused on the design of on both a 250 ton/day (solid feedstock) gasification system and a 500 ton/day gasification system with all the downstream systems necessary for syngas cleaning and conditioning. The resulting clean syngas will be capable of being fed dirctly to a catalysis plant at the proper H2:CO ratios required for liquid fuel production. As part of this effort, Emery is evaluating the variables that will affect the syngas cleaning and conditioning as a function of the feedstock inputs and the downstream catalysis process requirements. The downstream catalysis processes considered include: mixed alcohols (with emphasis on ethanol); Fischer Tropsch liquids; and methanol to DME. Each product option will affect the final gasifier configuration and the extent of post gasification conditioning equipment. These preliminary scales represent an initial deployment size that will be followed by subsequent scaling toward 1000 and 3000 ton/day systems.

655d Two Phase Anaerobic Treatment Process for Hydrogen and Electricity Production from Biodiesel Glycerol Waste

Baris Calli, Karolien Vanbroekhoven, and Ludo Diels. Environmental and Process Technology Department, VITO, Flemish Institute of Technological Research, Mol, Belgium

This paper discusses a two phase energy efficient anaerobic process composed of dark fermentation and biologically catalyzed electrolysis to treat the biodiesel glycerol waste as well as produce hydrogen and/or electricity. Biodiesel is a promising renewable fuel which will presumably be produced more intensively in the coming years. Glycerol is the major by-product of biodiesel production however it is highly contaminated with salts formed during neutralization of the alkali used in transesterification process. For every 10 litre biodiesel produced of about 1 kg glycerol is generated. The rapid increase in biodiesel production will result in a great surplus of glycerol in near future. Besides, glycerol derived from animal fats and frying oils would not be favourable as a commercial product. Therefore biodiesel glycerol seems to rapidly become a waste material that should be managed in an energy efficient way. Although the generation of hydrogen through glycerol fermentation is known for decades, researches and practices on dark fermentative hydrogen production from biodiesel-glycerol waste are very limited. Another novel and energy efficient way to manage biodiesel glycerol waste is production of hydrogen and/or electricity using a bio-electrochemical process such as biologically catalyzed electrolysis. In a bio-electrolyser, hydrogen is produced through bacterial electrolysis of glycerol by only applying an extra electrical energy about 0.25V which is substantially lower than the energy (~1.5V) necessary for conventional water electrolysis. Although, both processes could be used separately, a two phase system composed of dark fermentation plus bio-electrolysis would provide higher energy yields and treatment efficiencies. Since, the two phase anaerobic system proposed here for the treatment of biodiesel glycerol waste present an innovative and promising technology, this study will blaze a trail in developing a new energy efficient treatment concept.

Dark H2 fermentation experiments are conducted in completely stirred fermenters inoculated with mixed culture compost materials. The H2 consuming methanogens in the inoculum are eliminated by keeping pH quite low at the start-up. To maintain the thermophilic conditions (55°C), fermenters are placed in hot water circulated water jackets and rigorously mixed with magnetic stirrers to maximize H2 stripping. To control the pH during fermentation, pH transmitters equipped with pH electrodes and pumps dosing NaOH and HCl are used. The biogas evolved is continuously measured using volumetric gas counters appropriate for small volumes and be instantly released from fermenter. The H2 gas is monitored on-line with H2 sensors and periodically with gas chromatograph analyses. To prevent decomposition in the feeding tank waste glycerol is kept at 4°C and fed to the fermenters with peristaltic pumps. Glycerol in the influent and effluent is analyzed with a HPLC equipped with an ELSD detector. VFAs, acetone, alcohols and lactate are analyzed using a GC-FID. The effluents of hydrogen fermenters are fed to the bio-electrolyser unit after appropriate dilutions.

Bio-electrolysis experiments are conducted in a two chamber electrochemical cell. Anode and cathode chambers are separated by a cation exchange membrane and both equipped with a Ag/AgCl reference electrode to measure the electrode potentials. Graphite felt is used as electrode material in anode and cathode and connected to a potentiostat via isolated electrical wires. The potentiostat is used to control the applied voltage on the total system (as the power supply) and to control the anode potential during enrichment of electrochemical bacteria after inoculation. The experiments are performed between 25 and 30°C. Current and applied voltage measurements are done by the potentiostat. Hydrogen and CO2 gases produced are monitored in the same way as in dark fermentation experiments.

Immobilized Biological Hydrogen Fermentation with Chloroform Treated Granule

Bo Hu[a], Shulin Chen[b]*
a. Department of Chemical Engineering, University of Puerto Rico, Mayaguez, PR, 00681-9046;
b. Department of Biological Systems Engineering, Washington State University, Pullman, WA, 99164-6120
* Corresponding author

Introduction

Biological hydrogen production has attracted more and more attention in recent years because of its ability to produce hydrogen energy while also addressing concerns on environment and fossil fuel replacement. In addition, the process is particularly environmental friendly because negatively valued materials (such as agricultural waste) can be used in the process. Hydrogen is an intermediate product during the anaerobic digestion. Different groups of microorganisms work together as a food chain in an anaerobic digestion system to degrade the organic compounds. Acidogenic bacteria degrade organics to produce volatile fatty acid (VFA) and hydrogen, which is called acidogenesis. Then, acetogenic bacteria produce acetate from VFA and solvents with acetogenesis. There is a group of acetogenic bacteria that can synthesis acetate from hydrogen and carbon dioxide, a process called homoacetogenesis. And finally methanogens use acetate or hydrogen to produce methane as the final product. Among these microorganism groups, bacteria species involved with acitogenesis and acetogenesis being identified include Clostridia species. Clostridia are Gram-positive, spore-forming and acid producing bacteria that can tolerate harsh environmental conditions by forming spores. To produce hydrogen as a final product, hydrogen consumers, such as methanogens, need to be eliminated from the system, which is called pretreatment process. Most studies on microbial hydrogen production have used suspended-cell systems, which are normally inefficient or difficult to control in continuous operations. The low growth rate of hydrogen producing bacteria limits the productivity of a suspended-growth reactor due to the requirement for long hydraulic resident time (HRT) to maintain adequate bacteria population. Different microbial immobilization methods can be used for bacteria retention, for example, calcium alginate entrapment. There are various challenges, however, for the polymer matrices to entrap hydrogen-producing bacteria during the continuous operation. Granulation, which immobilizes the bacteria by self-flocculation, is simple to use and easy to control at an industrial scale. This immobilization technique can be used in biological hydrogen production. It was found that hydrogen producing biomass can develop into granules with high bioactivity. Hydrogen producing sludge was first demonstrated to agglutinate into granules in a well-mixed reactor treating a synthetic sucrose-containing wastewater at 26ºC, pH 5.5, with 6 hours of hydraulic retention. The granules became visible by day 15 and the size stabilized at around 1.6 mm diameter after day 60 (Fang *et al.*, 2002). Rapid and efficient granular sludge formation was found with Carrier-Induced Granular Sludge Bed (CIGSB) bioreactor with addition of support carriers, especially activated carbon (Lee *et al.*, 2004). Overall, without the carrier inducement, direct granulation of hydrogen-producing bacteria is a time consuming process and there are many unknown factors that need to be investigated during the process. This paper explored a creative bacteria immobilization method for hydrogen fermentation. Anaerobic granular sludge was proposed as the immobilized hydrogen producing bacteria to be used in hydrogen fermentation after methanogenic activity of the granules was eliminated in the pretreatment process with the chloroform. Immobilized hydrogen production with chloroform treated granule was investigated in this study, and then the biological hydrogen production was combined with methane production for efficient use of VFA. The microbial community changes inside the granules during the chloroform treatment and continuous culture were also investigated in this study. Culture-independent method, the analysis of terminal restriction fragment length polymorphisms (TRFLP) of 16S ribosomal RNA (rRNA) was used first to analyze different samples in each condition. When comparing the TRFLP data generated from different communities, these data provide quantitative information on the compositional differences of granular microbial communities. As long as the community composition is consistent for each condition, only representative samples in each condition were chosen to build the clone libraries for detailed description of community composition.

Materials and methods

Seed sewage sludge and methanogenic granules

The anaerobic sewage sludge was taken from the Wastewater Treatment Plant in Pullman, WA. The methanogenic granules were taken from a UASB treating starch wastewater, and the average settling velocity was 29.5 m/h (measured with 1000 ml graduated cylinder full of DI water, 100 samples).

Chloroform pretreatment

20 ml of culture medium was cultured with 0.25% chloroform addition for over one day, then, the granules were filtered and washed for inoculation.

Glucose culture medium

The medium used for H_2 fermentation contained 20 g glucose COD /L (i.e., 18.75 g glucose /L) as the carbon source and sufficient amounts of inorganic supplements, including: NH_4HCO_3 (5.24g/L), $NaHCO_3$ (6.72g/L), K_2HPO_4 (0.125g/L), $MgCl_2 \cdot 6H_2O$ (0.1g/L), $MnSO_4 \cdot 6H_2O$ (0.015g/L), $FeSO_4 \cdot 7H_2O$ (0.025g/L), $CuSO4 \cdot 5H_2O$ (0.005g/L), and $CoCl_2 \cdot 5H_2O$ (1.25×10^{-4}g/L).

Batch culture

Twenty ml of culture medium was placed in a serum bottle with three ml pretreated sewage sludge or pretreated granules. Nitrogen gas was pumped into the serum bottle for 2 minutes before the fermentation to eliminate the oxygen inside the bottle. The culture temperature was 35°C and the shaking speed was 100-150 rpm (Incubator IC 600, Yamato).

Repeat culture

After batch culture of granules with chloroform addition, another batch culture was performed without chloroform addition to see whether chloroform pretreatment caused irreversible damage to hydrogen producing bacteria and methanogen. The granules were taken out, washed with water and filtered. These granules were put into a serum bottle with the same amount of original culture medium to start another batch that was cultured with the same conditions as described above. Also, this procedure was repeated for several times to determine if the chloroform treated granules can be reused.

Continuous culture

An upflow reactor was set up with a volume of 450 ml and the height of the reactor was 18 cm. 50 ml methanogenic granules were treated with 0.1% chloroform following the method mentioned previously. Chloroform treated granules were packed in the upflow reactor as an expanded bed. Glucose culture medium 20g COD/L was fed into the reactor at the bottom with the pump. Biogas and fermentation broth flowed out of the reactor at the top and were separated in a separation bottle. Water replacement bottles were connected with separation bottle to collect the biogas and measure the biogas volume. The reactor was maintained at batch mode for three days, and then began continuous mode at Hydraulic Retention Time (HRT) = 13 hours. The HRT was adjusted quickly to 5.3 hours and kept operational for three days until it ran constantly (constantly here means that the pH of the effluent, the biogas and hydrogen production, the glucose conversion and VFA production are constant for over 12 hours).

Analysis

Total biogas production was measured at the end of the batch fermentation. The biogas was released into a U-tube with water. The biogas volume produced during the fermentation was measured through the water pressured out of the U-tube. The composition of biogas (H_2, CO_2, CH_4 and H_2S) in the headspace of the reactor was measured using a gas chromatograph (GC, CP-3800, Varian, Walnut Creek, CA) equipped with detectors including a thermal conductivity detector for H_2 and CO_2, a flame ionization detector for CH_4, a Valco Instrument Pulsed Discharge Detector run in Helium Ionization Mode D2 for H_2S, an 18' × 1/8" HayeSep Q 80/100 Mesh Silcosteel column for CO_2, H_2 and CH_4 with nitrogen as the carrier gas, and a 50m × 0.53mm × 4µm SilicaPLOT column for H_2S with helium as the carrier gas. Volatile Fatty Acids (VFA) were analyzed using a Dionex DX-500 system (Sunnyvale, CA, USA) containing an AS11-HC (4mm 10-32) column, a quaternary gradient pump (GP40), a CD20 conductivity detector, and an AS3500 auto-sampler. Settling velocity of granules was measured in water.

Sample preparation for 16s rDNA microbial analysis

Condition 1 was the methanogenic granules, taken from a UASB treating starch wastewater. Condition 2 was the chloroform treated granules after 5 days culture and Condition 3 was the samples taken from the upflow reactor after 15 days culture of chloroform treated granules.

DNA extraction

Genomic DNA was isolated with a commercial kit (QIAamp, Qiagen, Valencia, CA), and resultant DNA was stored at 4°C. The concentration of the genomic DNA was measured by both OD 260 and electrophoresis in 1.5 % agarose gel followed by staining with ethidium bromide and visualization under ultraviolet light.

PCR amplification

Internal fragments of 16s rRNA sequences were amplified from genomic DNA (gDNA) using universal bacterial primers 8f and 926r (based on *Escherichia coli* positions) (Zhou *et al.*, 2004), and universal archaea primers 112f and 934r separately (Baker *et al.*, 2003). Amplification was performed in 50 µl (total volume) reaction volumes that contained 20 ng (1 ul) of sample DNA, 1U of AmpliTaq DNA polymerase (Roche Molecular System, Inc., Branchburg, New, Jersey, USA), 1 x AmpliTaq reaction buffer, 1.5 mM $MgCl_2$, 100 mM dNTP, 5% DMSO and 0.05 mM each primer (Applied Biosystems, Middletown, CT). Initial DNA denaturation at 94°C in a PTC-100 Programmable Thermal Controller (MJ Research, Inc.) for 5 min followed by 35 cycles of denaturation at 94°C for 1 min, annealing at 55°C (bacteria primers) or 60°C (archaea primers) for 1 min, and elongation at 72°C for 2 min, which was followed by a final extension at 72°C for 10 min. To confirm amplicon production, the mixture was analyzed by electrophoresis in 1.5 % agarose gel followed by staining with ethidium bromide and visualization under ultraviolet light.

Terminal restriction fragment length polymorphisms (T-RFLP) analysis

Fluorescent labeled primers (see above) were used to amplify gDNA to obtain a PCR product. C.a. 40 ng – 60 ng of each PCR product was digested at 37ºC for 3h with 10 U restriction enzymes HaeIII and MspI separately with enzyme-specific buffer. This mixture was then incubated for 20 min at 65°C to inactivate enzymes. This product (1µl) was then mixed with 0.5 µl of GeneScan-500 ROX size standard (Applied Biosystems) and 10 µl of formamide, then denatured at 95ºC for 5 min and immediately placed on ice for at least 5 min. The T-RFLP analysis was performed on the ABI PRISM 3100 Genetic Analyzer (3100 PRISM Genetic Analyzer, Applied 150 Biosystems).

16s rRNA gene clone library

A clone library from the amplified DNA was generated with the TOPO TA Cloning kit. C.a., 20ng fresh 16s rRNA PCR product were cloned into PCR2.1-TOPO vector (20 ng) (Invitrogen Corporation, Carlsbad, NM, USA) using a vector/insert ratio of 1:1 and procedures recommended by the manufacturer. TOPO cloning reaction products (3 µl) were transformed into a vial of One Shot Chemical Competent *E.coli* (Invitrogen Corp) that were subsequently plated onto Luria-Bertani-Hogness agar plates containing 100 mg/ml kanamycin and 20 mg/ml x-gal to verify blue/white selection and incubated overnight at 37°C. Approximately 96 white, well-isolated white colonies were randomly selected from each library and grown in 1.5 ml of Luria-Bertani broth containing 100 mg/ml kanamycin in 96-well micro liter plates for 48 h. These cells were then harvested and the plasmid DNA was isolated using QIAprep96 Turbo Miniprep Kits (QIAGEN Inc) using a QIAGEN BioRobot 3000 workstation.

Sequencing and Sequence Analysis

Approximately 600 isolated plasmids with cloned inserts were sequenced with both bacteria 926r and archaea 934r primers. The insert sequences were determined using Big Dye ver. 3 cycle sequencing reactions (Applied Biosystems), and resolved using an automatic sequencer (3100 PRISM Genetic Analyzer, Applied 150 Biosystems). Sequences were trimmed to exclude the PCR primer binding sites and manually corrected with Chromas 2 (Chromas Version 2. 22; http://www.technelysium.com.au/chromas.html). For identification of closest relatives, insert sequences were compared to those available in the Ribosomal Database Project [RDP] and GenBank databases [http://www.ncbi.nlm.nih.gov/] using standard nucleotide-nucleotide BLAST program [blastn]; [http://www.ncbi.nlm.nih.gov/].

Results and discussions

Chloroform pretreatment

Chloroform treatment has been found in our previous study to be an effective method for eliminating methanogenic activity and for switching a methane producing system to hydrogen producing system. Chloroform selectively inhibits the methanogenic activity while allowing normal hydrogen production, as long as chloroform concentrations are low. Chloroform, however, causes damage to methanogens and the methanogenic activity is not easily recovered even upon removal of the chloroform in subsequent cultures. In this study, the low concentration of chloroform added in the culture medium in the pretreatment did not cause any damage to the structure of the granules. However, the chloroform treatment dramatically changed the microorganism distribution after several days of culture. Filamentous methaongens are clearly visible on the surface of the untreated granules, but there are no visible filamentous microorganisms on the chloroform treated granules in the SEM picture. Similarly, in the TEM picture, the methanogens are distributed and sectioned from the methanogenic granules, while the TEM picture of the chloroform treated granules is much clearer without the methanogens. Chloroform showed good selectivity on inhibition of hydrogen-producing bacteria and methanogens, eliminating the methane production while allowing normal hydrogen production.

Influence of HRT on the continuous granulized fermentation

The productivity reached 11.6 L/L/day at HRT 5.4 hours of continuous hydrogen production in the upflow reactor packed with chloroform treated granules. The granules changed their color gradually from deep black to dark grey, which might be caused by the washing out of the sulfur reducing bacteria. The granular structure was mostly kept after 15 days operation. New granules, which have shallow grey color, could be seen after ten days of operation. Long- term continuous operation was continued to investigate the factors influencing the upflow reactor. As HRT decreased, hydrogen productivity gradually increased. The yield of hydrogen fermentation, however, decreased when the HTR was shorter than three hours, owing to the overloading of the upflow reactor.

Influence of initial pH of the culture medium on the continuous granulized fermentation

For the continuous fermentation in the upflow reactor, hydrogen production decreased when the initial pH of the culture medium was below 6.0. In our previous batch study the culture conditions appeared to be too harsh for the hydrogen-producing bacteria to grow; nearly no hydrogen or volatile fatty acid (VFA) was produced when the initial pH was below 4.0 or above 9.0. The previous batch study also showed that the initial working pH for the batch culture medium ranged from 5.0 to 8.0, and that hydrogen production did not show a significant change within this initial pH range (Hu *et al.*, 2007). It seems that continuous culture in the upflow reactor required a narrower initial pH range of the culture medium. There was no significant difference for the hydrogen productivity between initial culture medium at pH 7.0 and 8.0, and the pH of the fermentation broth at these two conditions was around 6.0. When the initial pH was 5.0 or 6.0, the pH of the fermentation broth out of the reactor was about 4.2. In this case, the granules inside the reactor were drifted up with the upflow and gradually washed out of the reactor, which drastically decreased overall hydrogen productivity of the system. It is possible that the fermentation broth was too acidic, and therefore, it might have been erosive to the extracellular polymer that maintains the granular structure. Furthermore, because hydrogen-producing bacteria grow very slowly in acidic conditions, there might have been more bacteria detaching from the granules than bacteria reproducing and attaching to the granules. The granules washing out of the reactor appeared more like floc without granular structure, which confirmed that the continuous granulized hydrogen production should remain in around neutral pH conditions because the granular structure collapsed in very acidic conditions.

Influence of glucose concentration of the culture medium on the continuous granulized fermentation

Substrate concentration influenced the continuous granulized fermentation. With an increase in glucose concentration and the COD loading of the upflow reactor, hydrogen productivity reached maximum levels and then decreased. There was no significant difference in hydrogen productivity between glucose concentration at 20 g COD/L and at 30 g COD/L. When the glucose concentration was over 20 g COD/L, however, the overall yield decreased gradually. Fermentation with high substrate concentrations are preferred because many raw materials (such as whey or manure) have high COD content, and high concentrations of substrate are easy to heat and handle. In this study, cultures with substrate concentration below 20 g COD/L have nearly the same yield; however, at higher substrate concentrations, overloading occurs, causing a decrease in hydrogen productivity and overall yield.

VFA production and integration with immobilized methane production

Biohydrogen production, coupled with the subsequent step of methane production, can be very efficient. After three days of biohydrogen fermentation with chloroform treated granules, the fermentation broth was inoculated with methanogenic (untreated) granules for methane production. The initial pH of the culture medium was 8.0 for the hydrogen production and the pH of the broth decreased to 6.5 due to volatile fatty acid (VFA) production before converting to a methane production reactor. Overall, the integration of the immobilized hydrogen production with immobilized methane production provided 1.01 mol hydrogen and 2 mol methane per mol glucose. It is well known that formation of VFA during acidogenesis of organic matter precedes methanogenesis. The fermentation broth of biohydrogen production can therefore be used with acetogensis and methanogensis to produce methane, which can alleviate the costs and environmental concerns associated with the biological hydrogen production process. Production of methane from hydrogen fermentation broth brings extra value by producing methane, and the degradation of VFA also decreases the overall COD value of discharge from the hydrogen production process. The conversion of biohydrogen production follows the reaction as:

$$C_6H_{12}O_6 + 2H_2O \rightarrow 4H_2 + 2CH_3COOH + 2CO_2$$

$$C_6H_{12}O_6 \rightarrow 2H_2 + CH_3CH_2CH_2COOH + 2CO_2$$

If we take acetate as the final product of biohydrogen production, the theoretical integration of biohydrogen production and methane production can be illustrated as:

$$C_6H_{12}O_6 + 2H_2O \rightarrow 4H_2 + 2CH_4 + 4CO_2$$

It is evident that hydrogen production can still be improved. In Stage 1, only 25% of hydrogen was produced (nearly 75% was unaccounted). The reasons are multifarious. For the batch culture, with an increase in partial hydrogen pressure of the head space, the pathway switches to butyrate production instead of acetate, which gives theoretically only 2 mole hydrogen per mole glucose. If the pathway goes to another end product, such as solvent or propionate, no hydrogen will be generated. Ways to further increase hydrogen production is still a topic of investigation for many researches, including the authors of this study.

Methane production reached 1.99 mol per mol glucose, nearly the same as the theoretical value of the integrated process. Because there are many other end products in Stage 1, such as butyrate, lactate, and so on, the missing hydrogen generation capability appears to be directed to methane production with acetogensis, which produces hydrogen and acetate from butyrate, propionate and solvent, and provides the necessary raw material for methanogensis.

There are thousands of UASB reactors running worldwide that can anaerobically degrade various organic wastes for methane production. This study provides a modification process for current granulized anaerobic digestion system by adding a separate upflow reactor packed with chloroform treated granules to harvest hydrogen before the waste stream feeds into the UASB. The integrated two-stage design for immobilized hydrogen fermentation and methane production offers a promising approach for modifying current anaerobic wastewater treatment processes to harvest hydrogen from the system.

T-RFLP analysis of microbial during the chloroform treatment

For the 16s rDNA based microbial analysis, microbial T-RFLP images showed similar peaks within each treatment, which indicates that there were limited differences in microbial communities within treatments, but we observed several differences between treatments. Bacteria T-RFLP image showed many peaks from samples of methanogenic granules, and fewer peaks from the samples of chloroform treated granules. Some new peaks showed on the image for the chloroform treated granules and the relevant abundance is different from the methanogenic granules. These data confirm that there is a bacterial community change during the chloroform treatment. T-RFLP image of samples from the upflow reactor showed the fewest peaks, but these appear to have corresponding peaks in both methanogenic and chloroform treated granule samples. This suggests that with the reactor, the community became less diverse relative to the other conditions tested herein.

Archaea T-RFLP image showed many peaks from the sample of the methanogenic granules. Interestingly, samples of chloroform treated granules retained most of these peaks despite an absence of methanogenic activity. This might be explained by the presence of gDNA in metabolically inactive methanogens, or non-specific amplication of non-archaea targets. To examine this question further, we generated a 16s rRNA clone library.

16s rRNA libraries

Over 400 base of the sequence of each clone was compared to Genebank. We assigned presumptive genus names if there was ≥ 90% sequence similarity to an existing sequence. Clones with ≥ 97% sequence similarity were presumed to be of the same species, and clones with <90% sequence similarity were designated as novel phylotypes. For the bacteria clone library, there were four major groups of bacteria in the methanogenic granule sample, including *Paenibacillus sp.* (13/86), *Enterococcus sp.* (7/86), *Clostridium sp.* (43/86) and *Bacteroides sp.* (16/86) (*Bacteroides sp.* and *Petrimonas sulfuriphila* are combined here owing to a close phylogenetic relationship). There were also four novel phylotypes in the community. After chloroform treatment, *Paenibacillus sp.* and *Bacteroides sp.* were reduced or eliminated from the granules, while hydrogen producing species *Eubacterium sp.* (22/95) and *Enterococcus sp.* (18/95) increased. For the sample from upflow reactor, both *Enterococcus sp.* and *Eubacterium sp.* disappeared after 15 days flushing. *Clostridium sp.*, which remained almost the same prevalence in the library following the chloroform treatment, dramatically increased inside the upflow reactor, this is consistent with a shift towards hydrogen production. The *Clostridium* species composition appeared to change with the chloroform treatment and continuous culture. *Clostridium butyricum*, a typical hydrogen producing species with butyrate pathway, became the dominant representative over time. With the increase of the hydrogen partial pressure, the pathway shifts to the butyricum pathway, and then a non-hydrogen producing pathway, such as formation of lactic acid. Thus, it was not surprising that we observed *Lactobacillus sp.* (23/94) in upflow reactor sample as this was consistent with the increasingly favored non-hydrogen pathway.

For the archaea clone library, most phylotypes were uncultured archaeon or euryarchaeote sequences. The uncultured euryarchaeote gene for 16S ribosomal RNA, partial sequence, clone:EA02; the uncultured archaeon TA05 16S ribosomal RNA gene, partial sequence and the uncultured archaeon TA01 16S ribosomal RNA gene, partial sequence were closely related to *Methanosaeta concilii* while the uncultured archaeon Arc No. 5 16S ribosomal RNA gene, partial sequence and others were close to other *Methanogen sp.* The archaea clone library for the sample from the upflow reactor included a number of clones of *Lactobacillus sp.* indicative of low selectivity from the archaea primers. This can also explain the difference of T-RFLP between chloroform treated granule sample and samples from upflow reactor. Ovcrall, the archaea clone library for chloroform treated granules and sample from upflow reactor were similar, while there was a significant archaea community change during the chloroform treatment. Most of methanogens disappeared after chloroform treatment, while *Methanosaeta concilii (Methanothrix soehngnii)* clones remained constant in the library. Most theories about granulation indicate that the acetotrophic methanogens *Methanosaeta concilii* (*Methanothrix soehngnii*) play a key role in the granulation (Pol *et al.*, 2004). Hydrogen producing system can form granules too. In our experiments, most of the *Methanogen sp.* was presumably eliminated during the chloroform treatment, consistent with the measured switch from methane producing system to hydrogen producing system. The presence of *Methanosaeta concilii* probably explains why granulation remained stable when the system produced mostly hydrogen (hydrogen producing bacteria are not known to produce granules). *Methanosaeta concilii*, which was renamed as *Methanosaeta concilii*, is a filamentous methanogen species that can only use acetate as its sole energy source and cannot use hydrogen as raw material. It is difficult to reconcile the presence of active methanogens in the absence of detectable methane, although if the total population remained very low methane production may have remained undetectable while PCR amplification of the target template might be sufficiently sensitive to amplify this genetic marker. Further research on the pathway of *Methanosaeta concilii* or its role in hydrogen producing granulation is needed.

Selected References

1. Fang, H. H. P., Li, Y. Y., and Chui, H. K. (1995) Uasb Treatment of Waste-Water with Concentrated Mixed Vfa. Journal of Environmental Engineering-Asce 121: 153-160.
2. Hu, B. and Chen, S. (2007) Pretreatment of methanogenic granules for immobilized hydrogen fermentation. International Journal of Hydrogen Energy In Press, Corrected Proof.
3. Lee, K. S., Wu, J. F., Lo, Y. S., Lo, Y. C., Lin, P. J., and Chang, J. S. (2004) Anaerobic hydrogen production with an efficient carrier-induced granular sludge bed bioreactor. Biotechnology and Bioengineering 87: 648-657.
4. Pol, L. W. H., Lopes, S. I. D., Lettinga, G., and Lens, P. N. L. (2004) Anaerobic sludge granulation. Water Research 38: 1376-1389.
5. Zhou, X., Bent, S. J., Schneider, M. G., Davis, C. C., Islam, M. R., and Forney, L. J. (2004) Characterization of vaginal microbial communities in adult healthy women using cultivation-independent methods. Microbiology-Sgm 150: 2565-2573.

655f Use of Gas Transfer Membranes for Recovery of Renewable Energy in Anaerobic Digesters

Nilesh K. Patel and Carlos D. M. Filipe. Chemical Engineering, McMaster University, 1280 Main Street West, Hamilton, ON L8S4L7, Canada

"Hydrogen energy" could be a possible sustainable energy source to reduce global emissions of green house gases if it is produced from renewable sources such as biomass and wastewater. Biological hydrogen production in anaerobic bioreactors can be increased by either producing more H_2 by fermentative bacteria and/or by eliminating H_2 consuming methanogens. Removal of the dissolved gases (H_2, CO_2, CH_4 etc.) from the liquid phase was accomplished by the use of hydrophobic microporous membranes. Reduction in liquid phase H_2 partial pressure decreases the free energy and more H_2 was produced by fermentative bacteria. H_2 consumption also decreased, as less substrate H_2 was available for H_2 consuming methanogens. Objective of present study was to test the gas transfer membranes that would stop H_2 consuming methanogenesis in anaerobic bioreactor.

A membrane contactor with polypropylene microporous hollow fibers was used for dissolved gas removal from liquid phase. Two bioreactors, 1st with membrane contactor and 2nd without contactor (control), with 2-L liquid volume and 1-L headspace were operated at 35 °C with hydraulic retention time of 5 d. Liquid from 1st bioreactor was circulated through membrane contactor to facilitate dissolved gas removal from liquid phase. The pH was controlled at 5.5. Glucose and nutrient solution were fed at the rate of 1:1 with organic loading rate of 3.0 gGlucose/L/d.

Volatile fatty acid analysis using gas chromatography (GC) showed clear shift in metabolic pathway. Acetate and butyrate were more than 10 times higher in the 2nd bioreactor (control) while propionate was more than double in 1st bioreactor (with membrane contactor). Chemical oxygen demand (COD) removals as a result of gas production were increased 3-4 times with using membrane contactors. These results indicated that more renewable energy in terms of gas production such as CH_4 and H_2 can be recovered by using gas transfer membranes. Results also showed that lower hydraulic retention time is required to stop H_2 consuming methanogenesis completely.

661a Removal And Recovery Of Solvents From Fermentation Broth By Pervaporation Using Liquid Membrane

Atsawin Thongsukmak and Kamalesh K. Sirkar. Chemical Engineering, New Jersey Institute of Technology, Membrane Seperations, 161 Warren st, Newark, NJ 07102

Pervaporation is an attractive technique for separation and recovery of volatile fermentation byproducts present at low concentrations due to its lower energy cost compared with distillation. Although polymeric rubbery membranes or ceramic membranes are quite successful in removing most volatile organic compounds (VOCs) from their dilute solutions in water, their selecitivies for removing polar organic compounds are not very high. Previous studies indicated that liquid membranes can achieve much higher selectivity for some small molecules such as acetic acid, ethanol, butanol, etc. In the current work, we have focused on liquid membranes immobilized in novel porous hollow fiber-based membranes as a support for the pervaporation process to increase the VOC selectivity and membrane stability. Volatile fermentation byproducts such as acetone, butanol and ethanol present in low concentrations in synthetic aqueous solutions have been separated by novel liquid membranes. We have tested the liquid membrane with model solutions (containing acetone-butanol-ethanol (ABE)) as well as a filtered solution from a fermentation broth (from ABE fermentation) by the pervaporation process. Very high selectivities for the solvents were obtained in the permeate especially at higher feed temperatures (> 50°C). The selectivities of butanol, acetone, and ethanol achieved were 275, 220, and 80 respectively with 11.0, 5.0, and 1.2 g/m2-hr for the mass fluxes of butanol, acetone and ethanol respectively at a temperature of 54 °C for a feed solution containing 1.5 wt % butanol, 0.8 wt % acetone, and 0.5 wt % ethanol. High stability of liquid membrane was also observed with virtually no contamination of the broth. Fluxes and selectivities of solvents were studied over a range of temperatures; higher temperatures yielded much higher selectivities. Thinner membranes have been developed in different thicknesses. The thickness of liquid membrane was reduced to 5 times thinner than the full substrate thickness, which yielded mass fluxes of solvents 5 times higher than those through full thickness liquid membrane. Thinner liquid membranes did not lead to a change in the selectivities at various temperatures. The thinner liquid membranes have been tested with model solutions and filtered fermentation broth from ABE fermentation at various temperatures.

661b Using Regional Biomass Processing To Co-Produce Animal Feed And Ethanol From Grasses

Bryan Bals[1], Hannah Murnen[1], **Bruce Dale**, Michael Allen[2], and David Main[2]. (1) Chemical Engineering and Material Science, Michigan State University, 2527 Engineering Building, East Lansing, MI 48824, (2) Animal Science, Michigan State University, 1290 Anthony Hall, East Lansing, MI 48824

A primary concern regarding the role of biomass in America's energy future is the debate regarding land devoted to food vs fuel. Critics claim that dedicated energy crops will reduce the amount of land available for human use, raising food prices or causing shortages. In addition, the large biorefineries needed to obtain process economies will require large tracts of farmland, making the logistics of the supply chain difficult. To address these issues, we propose a regional approach to biomass treatment, which can be used to enhance the value of biomass for both animal feed and ethanol production. Thus, land can be used for feed and fuel simultaneously, and the logistical supply problems can be significantly reduced.

In this study, we use ammonia fiber expansion (AFEX) pretreatment in conjunction with alkaline extraction and subsequent coagulation of proteins to co-produce a highly digestible lignocellulosic product and a high quality rumen bypass protein product from orchardgrass (Dactylis glomerata L., Baridana cultivar) and switchgrass (Panicum virgatum, Cave-in-rock cultivar). Protein extraction conditions (temperature, pH, residence time) will be optimized for both types of grasses, as well as the effect of additional additives such as surfactants. In addition, the effect of drying the grass, as well as the effect of AFEX pretreatment, will be studied in relation to the amount of protein obtained. Further protein processing may be necessary to decrease the degradation of protein within the rumen, thus enhancing the quality and therefore value of the final product. Heat coagulation in the presence of reducing sugars will be used to simultaneously concentrate and treat the protein extract, and the quality of the final product will be determined by both the extent of degradation and the amino acid profiles. Finally, a complete mass balance will be performed for the optimal processes for both grasses, indicating the effectiveness of the concept.

Web Page: www.everythingbiomass.org

661c Towards An Understanding Of Ionic Liquid Pretreatment Process Of Lignocellulosic Biomass: Recovery Of Ionic Liquid And Analysis Of Process Streams

Anantharam P. Dadi[1], Noureen Faizee[2], Kripa Rao[1], Sasidhar Varanasi[1], Constance A. Schall[1], and Jared L. Anderson[3]. (1) Department of Chemical & Environmental Engineering, The University of Toledo, 2801 West Bancroft Street, Toledo, OH 43606, (2) Department of Chemical & Environmental Engineerig, The University of Toledo, 2801 West Bancroft Street, Toledo, OH 43606, (3) Department of Chemistry, The University of Toledo, 2801 W bancroft street, Toledo, OH 43606

Ionic liquids being non volatile, non-flammable, recyclable and designer friendly, are gaining wide recognition as green solvents and efficient reaction media in process industries as a potential replacement for volatile organic solvents. Unique solvating properties of ionic liquids have made them good solvents for cellulose and lignocellulosic substrates. During ionic liquid pretreatment process, the recalcitrance of biomass due to high crystallinity of cellulose and lignin sheathing was overcome leading to high yields of glucan and xylan conversions. Rates of enzymatic hydrolysis for pure cellulose pretreated by ionic liquids were found to be two orders of magnitude higher than that for untreated cellulose [1, 2]. Enhanced enzymatic hydrolysis yields of glucan to glucose and xylan to xylose were observed for ionic liquid pretreated ligocellulosic substrates like corn stover and poplar reaching theoretical yields for glucan. However, during pretreatment process of lignocellulosic biomass, various process interactions take place leading to alteration of biomass structure and generation of various desirable or undesirable compounds. Analysis of the pretreated biomass and the various compounds generated will give a proper insight towards a fundamental understanding of the ionic liquid pretreatment process. Besides that an efficient recovery of ionic liquid is necessary which will make the ionic liquid pretreatment process economically viable. This work focuses on understanding the ionic pretreatment process by analysis of various process streams and addresses the recovery of ionic liquids by different approaches.

1 Anantharam P Dadi, P., Sasidhar. Varanasi, and Constance .A. Schall, Enhancement of cellulose saccharification kinetics using an ionic liquid pretreatment step. Biotechnol. and Bioengg, 95 (5) 2006 904-910

2 Anantharam P Dadi, P., Sasidhar. Varanasi, and Constance .A. Schall, Mitigation of cellulose recalcitrance to Enzymatic hydrolysis by ionic liquid pretreatment. Applied Biotechnol. and Bioengg 136-140 2007 407-421

661d Production and Hydrolysis of Electrospun Cellulose Fibers

Heidi J. Park, John W. Dingee, Sean R. Fitzgibbon, A. Brad Anton, and Yong Lak Joo. School of Chemical and Biomolecular Engineering, Cornell University, Olin Hall 120, Ithaca, NY 14850

The enzymatic conversion of cellulose to fermentable sugars is an attractive way to produce renewable feedstocks for fuels and chemicals; however, native cellulose is recalcitrant to enzymatic degradation and must be chemically and/or physically pretreated to improve its susceptibility to enzymatic attack. Previous hydrolysis studies using "model" cellulosic substrates have been unable to completely decouple the effects of various substrate properties such as degree of polymerization, degree of crystallinity, and accessible surface area on hydrolyzability; consequently, pretreatment recipes must be optimized largely by trial-and-error. Here, we *electrospin* cellulose to produce uniform fibers with carefully controlled and systematically varied microstructure to investigate the effects of various substrate properties on hydrolyzability. Knowledge gleaned from hydrolysis studies with electrospun cellulose has the potential to provide new directions for pretreatment that will optimize the hydrolysis of native cellulose. We present results for the effects of substrate properties on hydrolysis, and we discuss some new insights about the hydrolysis process that have been revealed through the use of electrospun cellulose as a model substrate.

661e Rheology Of Biomass Slurries

C. Tim Scott[1], Max Ehrhardt[2], Jianghui Wang[2], Thatcher Root[2], and Dan Klingenberg[2]. (1) USDA Forest Products Laboratory, One Gifford Pinchot Drive, Madison, WI 53706, (2) Chemical and Biological Engineering, University of Wisconsin, 1415 Engineering Drive, Madison, WI 53706

The conversion of biomass to energy is now more promising due to recent advances in the enzymatic hydrolysis of cellulose with subsequent conversion to ethanol. One factor in achieving economic viability will be processing the biomass at high solids concentration, so that the energy required for heating and other processing steps can be minimized, and the concentration of ethanol produced can be increased. However, processing at high concentrations creates challenges for pumping and mixing. While reducing particle size can improve the rheology, with added benefits for reaction kinetics, particle size reduction incurs additional costs. It is therefore important to understand the rheological and mass transfer properties of these slurries in order to design optimal, efficient processes. In this presentation, we report experiments employing various lignocellulosic materials, such as wood pulp and corn stover. Methods for measuring and controlling the particle size distributions will be discussed. Rheological measurements employ torque rheometry, a method commonly employed in the food and polymer industries. We describe the rheology of biomass slurries and its dependence on such variables as solids concentration, deformation rate, particle size and size distribution, and temperature.

661f Ethanol Enrichment by Corn Grit Adsorbents

Youngmi Kim[1], Nathan Mosier[2], Rick Hendrickson[1], Yifei Zang[1], Ahmad Hilaly[3], Joe Beggin[3], Kyle Beery[3], and Michael Ladisch[1]. (1) Laboratory of Renewable Resources Engineering (LORRE), Purdue University, Potter Engineering Center, 500 Central Drive, West Lafayette, IN 47907-2022, (2) Agricultural and Biological Engineering & Laboratory of Renewable Resources Engineering (LORRE), Purdue University, Potter Engineering Center, 500 Central Drive, West Lafayette, IN 47907-2022, (3) James R. Randall Research Center, Archer Daniels Midland Co., 1001 North Brush College Road, Decatur, IL 62521

Fixed beds of corn grits have been used to produce anhydrous fuel-grade ethanol in commercial scale ethanol industry over 20 years. Starch in the corn grits selectively adsorbs water by forming hydrogen bonds between free-OH groups on the glucose units of the starch and water molecules. The bed is regenerated by counter-currently flowing hot CO2 gas through the bed. The regenerating gas carries the desorbed water and ethanol left in the void volumes of the bed to a recycling system. An automated research scale corn grit adsorber was constructed for evaluating conditions for drying of an azeotropic ethanol/water mixture. We present results for breakthrough and cyclic operation of two different sizes of corn grits in terms of their adsorption capacities and product ethanol purity at conditions that simulate an industry scale operation. This paper also reports and compares general properties of fresh corn grits compared to corn grits that have been in operation for a year. Both were analyzed for changes in compositional and physical characteristics, which were found to be small. A fundamental understanding of the properties and characteristics of corn grits and their performance in a fixed bed mode provides knowledge for improving adsorbent lifetime and design of better-performing starch-based adsorbents.

AUTHOR INDEX

AUTHOR INDEX

AUTHOR INDEX

AUTHOR INDEX